Did Man Really Walk On the [Moon]
or was it the Ultimate Came[ra...]

Mystery Solved

DISCLAIMER:

This book is a controversial documentary questioning the authenticity NASA's alleged Apollo Moon landings back in the 1960's and 1970's. This book was written with the intention of examining evidence and arguments supporting the conspiracy theory that the Moon Landings were staged on Earth. All claims of misdoing contained in this book should be considered with the speculations and the Author's opinions. Although the theories expressed in this book appear solid, they may not be the only possible explanation and readers are encouraged to make their own judgments based on all available information.

All images unless specifically noted have been sourced from NASA's Apollo Mission archives. These images have not been altered except to adjust the size of the image, or cut out a section to fit the page. In some cases for illustration purposes a small section of an image has been magnified to help reveal certain problems.

At the time this book was written all images were available freely to the public on NASA's websites. Readers are encouraged to download the images in order to make your own conclusions.

The contents in this book are copyrighted © GTI Publishing all rights reserved

How America Faked The Moon Landings

Author, Charles T. Hawkins

How America Faked The Moon Landings
Author, Charles T. Hawkins

PUBLISHED BY GTI PUBLISHING
P.O. Box 0179, New Port. MN 55055

Book Title Home Page: Http;//www.moonbloopers.com
Authors Home Page: Http;//www.charleshawkins.com

Copyright by- Charles T. Hawkins 2004

This book is in copyright. Subject to statutory exception and to the provisions of relevant collection licensing agreements, no reproduction of any part may take place without the written permission of GTI Publishing.

Printed in the United States

Typeface 10/12 pt System Quarkexpress-TM

Hawkins, Charles -1962 A skeptical scientist who evaluates the authenticity of the Apollo Moon Landings and determined them to be fake.

How America Faked the Moon Landings - Did Man Really Walk On the Moon or was it the Ultimate Government Conspiracy / Charles T. Hawkins

ISBN - 0-9749405-4-2 Paperback

Index

Chapter 1: The Moon Hoax Conspirators- page 1-1

* A quest for the truth
* Neighborhood kids discover earth creature in moon landing film.
* Scientists from around the world join together to prove if the Americans Apollo Moon landings were real or not
* A challenge for NASA to start Answering Questions about their claims
* Cash **reward** for additional evidence ignites worldwide scavenger hunt
* Author's Very Unique Credentials

Chapter 2: The Conspiracy Theory- page 2-1

* Was the American Government capable of such a hoax?
* NASA hired a movie producer to oversee the entire Apollo Moon Landing project.
* Background of so-called crazy hoax conspirators is the opposite of what NASA wants everyone to believe.
* The way other countries really feel about the alleged Moon landings.
* Answer to why man hasn't been back to the moon.
* Earth Radiation Belt proves Moon Landings were fake.
* How NASA prevented deep space travel by exploding nuclear bombs in outer space.
* The Space Suits provided inadequate protection, therefore, the astronauts would have been burnt to crisp.

Chapter 3 - Suspicious Deaths - 3-1

Chapter 4: Basic Scientific Principles Prove Moon-landings were FAKE -page 4-1

* Suspicious deaths related to alleged Moon Landings
* Evidence suggests President John F. Kennedy was murdered over the Moon Landing Conspiracy
* Are American Teachers being forced by the Government to remain silent?
* Is NASA censoring media coverage of the Moon Landing event?
* The odds for successful Trip to the the moon and back was less than 1 in 6,000
* Whiz-kids Experiment #1 "Penetrating the Earth's Radiation Belts".

Chapter 5: NASA Builds Movie Studios And Calls Them Training Simulators- page 5-1

* How all the pieces of the Moon Landing hoax puzzle were put together.
* What started the Moon Hoax?
* How NASA training simulators were used to fake the moon landings
* Proof that training simulator photography taken at the Kennedy Space Center matches the ones allegedly taken on moon back in the 1960's and 1970's
* Earth based training simulator rock formations are almost identical to those in moon pictures.
* NASA imagery illustrates surface plumbing pipes camouflaged on the "moon surface" simulation
* The discovery of secret messages left by former NASA employees as clues for others to find

Chapter 6: How NASA Simulated the weightlessness of Space - page 6-1

* Mislabeled films show Moon Landings were staged on earth.
* Evidence shows how NASA artificially simulated weightlessness of the moon.
* NASA designed special equipment to assist the astronauts in filming the movie.
* How NASA used Look-A-Like Dummy Astronauts to make the fake Moon-landing photography.

Chapter 7: How NASA Faked the Mission Equipment Capabilities- page 7-1

* Proof NASA greatly exaggerated the spacecraft's capabilities.
* Proof the Lunar Lander Spacecraft did not fly under it own power as NASA claimed.
* Evidence reveals NASA had two groups of Apollo astronauts, one of real test pilots and one of actors. The real test pilots were all killed.
* NO Space equipment has been found left behind on moon surface as NASA claimed there were.
* Whiz kids Experiment #2: An independent Apollo spacecraft radiation level test would prove the moon landings were faked.
* Whiz kids Experiment #3 Using the Hubble telescope to prove the Apollo moon landings were faked.
* Whiz kids Experiment # 4 Using the Hubble telescope to prove the Apollo spacecraft was actually left in Earth's orbit, or returned to earth.

Chapter 8: NASA's Moon Landing Studios- page 8-1

* Evidence of no spacecraft Blast Crater, helps proves moon landings were faked.
* Time sequence photos reveals exactly how the Apollo moon landing movie stages were constructed.
* The theory that volcanic ash was used as a substitute for a Moon surface is confirmed by no tire tracks or footprints in many moon surface pictures.

Chapter 9: Proof the Moon Landing Photography Is Fake- page 9-1

* Proof not one of the moon landing photos is legitimate.
* NASA's custom modifications to the cameras used by the astronauts helps prove the photos were not taken on the moon.
* Whiz-kids experiment # 5: Proving the film should have melted in the astronaut's camera on the moon surface.

Chapter 10: Proof Stage Lights Were Used To Produce the Moon Landing Photography - page 10-1

* Photos are undoubtedly studio quality images that used special lighting equipment and retouching techniques
* Proof lights on the alleged moon surface match those lights at NASA's training simulators here on earth
* There are diverging shadows in many of the moon surface photos
* Pictures show how the stage lighting had created "Hot Spots' on the photos and what techniques NASA used to remove them from the pictures
* Moon Landing Photographs reveal the actual Stage lights on alleged moon landing sites
* Whiz-Kids Experiment #6 "HOT SPOT Confirmation Experiment:

Chapter 11: NASA Used Astronaut Training Photos- page 11-1

* Many Apollo astronauts training pictures match the photography that NASA claims was taken on the moon.
* Moon appearance created by a Day to Night Filter.
* Picture taken outside the Kennedy Space Center also match those alleged to have been taken on the moon.

Chapter 12: How NASA Created the Moon Scenery - page 12-1

* Solid evidence of background scenery paintings are used to fake moon landscaping
* Sliding Scenery photography trickery used to fake the moon's mountain background
* Composite Photography trickery was used to fake foreground scenery
* Artificial Moon Surface found at Langley Research Center matches that on moon
* Mountains mysteriously disappear on moon
* Apollo Mission Landing Site matches each other when superimposed

Chapter 13: The Earth Locations That Match Landing Sites on the Moon - page 13-1

* Special program developed pinpoints where moon landings were filmed here on earth

*Location of Matching Moon landing scenery here on Earth:

    1. The Rio Grande Gorge between Pilar and Taos, New Mexico
    2. Pancake Range area of south-central Nevada
    3. Coso Range area, near Ridgecrest, California
    4. Kilauea Volcano, Hawaii
    5. Pinacate Volcanic area of northwestern Sonora,
    6. White Sands Mexico
    7. Cinder Lakes,

Chapter 14: How NASA Created the Moon Craters - page 14-1

* NASA filmmakers used Man-made Craters
* NASA filmmakers constructed artificial craters from pieces of several different sceneries
* NASA filmmakers used actual earth Craters as moon craters

Chapter 15: Deciphering NASA's Coded System For Hiding Their Incriminating Evidence- page 15-1

* **Introduction to the "Amitch System"** (Artificial Machine Intelligence Technology Computerized Hyperthinking)
* The Whiz-kids from Japan provide a solution to major computing problem
* NASA's Apollo Mission photo numbering system reveals many clues to the moon landing hoax
* Very unusual items were discovered on the moon surface.

Chapter 16: How NASA's Moon Landing Movie Stages Functioned - page 16-1

* Details on How NASA's Moon landing movie studios functioned.
* NASA photographers simply combined Apollo training Photos to create the alleged moon landing photos
* Footprints and Tire tracks missing in the moon landing photos
* More proof rocks on the moon are labeled with English Language letters and numbers.
* More proof that artificial moon dust was used in the moon landing pictures.
* Moon rover photos from the moon confirm moon landing was a hoax
* Movie Stage Markers found on the moon.
* Pre-made flag pole holes are discovered on the moon surface.

Chapter 17: NASA Employees Leave Clues to the Moon Landing Hoax - page 17-1

Chapter 18: The Moon's Animal Kingdom - page 18-1

Chapter 19: Advanced Theories - page 19-1

* The Moon photos of Earth are totally inaccurate and out of place.
* NASA's own evidence proves the moon transmissions were really from earth, not the moon
* Astronauts Moon Surface Vacuum Test proves the Apollo moon landings were filmed here on earth
* The Communication Link NASA claims the Apollo Astronauts used could not have functioned
* NASA's own Solar Wind Test results will confirm the moon landings never took place.
* Custom Light Filters Simulate Moon's Dark Environment

Chapter 20: Whistle-Blower Clues - page 20-1

*NASA employees leave intentional clues to the Moon Landings being faked in hope that someday they would be found.

CHAPTER 1
The Moon Hoax Conspirators

TO BELIEVE, OR NOT TO BELIEVE, THAT'S THE QUESTION

CHAPTER 1
The Moon Hoax Conspirators

A quest for the truth...

Without a doubt in the 1950's and 60's the United States was faced with one of their greatest challenges of all time. Confidence in the military's ability to protect the nation was at an all time low. Nearly everyone believed it was inevitable that the Soviet Union would attack with nuclear bombs.

Schools routinely performed nuclear bomb drills and young children on playgrounds were often found chanting, "The Russians are coming, the Russians are coming". These fears of an inevitable nuclear attack were very real and played a major role in the American psyche. Many people out of fear were building bomb shelters in their homes and backyards, hoping that their family would somehow survive the nuclear attack they thought was certain to take place soon. Clearly something needed to be done to show the world that the United States still maintained military superiority and to calm people's nerves. The United States eventually established themselves as the superior technological power in the world by winning the space race and landing the first men on the Moon.

This book is based on a true story about a special group of Internet Whiz Kids who were convinced the United States Apollo Moon Landing project was a hoax. Filled with government cover-ups, corruption and murder to conceal the truth with evidence swept under the carpet, NASA did all they could to deceive the rest of the world. Yes, these Whiz Kids were convinced that the greatest achievement of mankind was actually the greatest hoax of all time. As strange as it sounds, this group thought that out of desperation the United States faked the Moon landings to obtain their technical advantage over the Former Soviet Union to win the space race. They believed the Apollo missions were merely elaborate marketing schemes and social hype, rather than legitimate engineering advancements and scientific breakthroughs.

Naturally I was skeptical at first, though as more and more evidence was uncovered, it appeared the Whiz Kids were on to something spectacular. They soon convinced me to help them in their quest to solve one of the greatest mysteries of all time.

This book details some of the most remarkable findings that support the Moon Landing Hoax theory and the amazing adventure we went through to make this book a reality. You'll find there is a great deal of never-before-seen evidence contained in this book suggesting the Moon Landings were faked here on Earth. Some of the evidence is very controversial. Please do not rely on your emotions, instead, honestly consider any evidence presented that may be contrary to your beliefs. Then after you have finished this book, if you are still not convinced the moon landings were fake, that's OK: refutability is one of the classic determinants of whether a theory can be called scientific. If nothing else, share this book and its information with your family and friends; you'll have the time of your life merely considering the possibilities and discussing them among yourselves.

While sharing this information with others, many of the Whiz Kids said they could see how easy it would have been for the Moon Landings to go unchallenged for so long. When they showed the evidence suggesting the Moon Landings were faked to their parents and teachers, the immediate response was of shock that someone would even suggest such a thing. Some people seemed to be conditioned to instinctively accept the words spoken by the "Almighty American Government", and refused to even look or listen to any evidence to the contrary. There is even a strong possibility that some of your friends will thank you for clearing their eyes. You may find sharing this information can be a lot of fun!

Although the evidence in this book may seem biased to the Moon Landings being faked, this was NOT the original intent. The goal has been to ensure this book contains the most accurate information ever compiled on paper related to the Apollo Moon Landing Missions, backed by scientific proof supporting any conclusions. NASA's evidence has always been given the benefit of doubt in confirming the Moon Landings were indeed real. The conclusion that a particular piece of NASA's Moon Landing evidence was indeed fake only resulted after all realistic alternatives of it taking place on the moon were eliminated. I was very skeptical at first, you know how some people are always claiming the American Government is doing something devious behind their back. Most claims have no basis and can easily be discredited scientifically. For instance, many people believe the chlorine that is added to the public drinking water is some type of mind controlling inhibitor the government is using to keep people semi-subdued at all times. Clearly, all scientific research done on chlorine proves this is not true.

The only thing the government is keeping secret about adding chlorine to the drinking water is that it's maybe one of the leading causes of heart disease, resulting in deposit build up in the artery walls. But in no way can a small amount of chlorine inhibit peoples thinking patterns. That's the beauty of science, it can in most cases confirm or disprove certain claims. Unfortunately, that is not always possible. For example, there is a well known urban myth that the CIA supplies drugs to inner-city black minority groups to promote poverty. This particular claim of CIA government intervention is obviously ridiculous. However, without any scientific evidence it is impossible to disprove or confirm this myth. Even though the government has worked hard to hide and cover up most of the information about the Moon landings, hard work and perseverance provided us with enough evidence to scientifically conclude whether or not the NASA Apollo Moon Landings were real.

How the "Whiz Kids Project" Started

One of the most common questions everyone seems to ask is, "How did this amazing quest to find the truth about the Moon Landings get started?" At first, it seemed unnecessary to mention the history of the book and dive right into all of the exciting Government Hoax conspiracy evidence. However, since knowing the background of the book will help establish the authenticity of the evidence discovered, it's necessary to briefly mention the topic.

The quest to solve the mystery began while I was watching TV one afternoon with my wife and our children. An important news flash showed film footage of the Apollo Astronaut, Buzz Aldrin, punching someone in the nose for asking him to swear on the Bible that he really went to the moon over thirty-four years ago. Our initial response was of shock. What would provoke one of the Apollo Moon astronauts to respond in such a way? No one would expect Buzz Aldrin to respond on demand for every request, he is most likely a very busy person. But to punch someone in the nose for asking that question seems a little strange to say the least.

My oldest boy turned and asked, "Dad, did we really go to the moon, what's up with that?" I replied, "Of course, the American Government said we did, they would never lie about something that big." Then I made the mistake of saying, "But that really strikes me as odd that someone who went to the moon would be above questioning." That immediately sparked a huge family debate, which set off a neighborhood debate with our children's friends and parents. We had more than fifty kids and parents trying to figure out if we really went to the moon. Many kids were digging up information on the Internet and some called their parents, grandparents, and even teachers. People were looking for information in encyclopedias and even going to the local library to find books on the Moon Landings. It was a scavenger hunt weekend using the power of modern technology.

To my surprise, the results were astonishing. The kids that used the Internet to collect their data were convinced without a doubt, that man had never been to the moon. The kids that used the public library and called their parents, grandparents, and teachers, all thought man went to the moon, but then, were later convinced to the contrary by the Internet WhizKids. Surprisingly, all of the adults and teachers were convinced the United States went because the American Government and NASA said they did.

Two things were very interesting about the Moon landing project. The first thing was how little anyone really knew about it. Second, it was disconcerting to know how closed minded the adults were for refusing to listen to any of the evidence the Internet WhizKids uncovered that indicated the moon landings were faked.

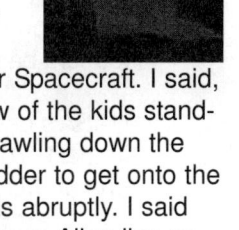

As Sunday night began to fall and everyone went home, our neighborhood adventure came to an end. You know how kids are: they usually move onto something else that catches their interest. The event soon faded from my memory as well. I did not give it much thought, until several weeks later, when I caught my kids and a couple of their friends gathered around the computer. "What are you guys up to?", I said, "How many times do I have to tell you not to download pirated movies on the Internet?" They quickly revealed they were downloading films of the Apollo Moon Landing from deep within the NASA archives." I told them they were wasting their time and stressed that the United States sent men to the moon. Pointing to the monitor they said, "Oh, you think so, well just look at this!" I saw what appeared to be a spider web on the Lunar Spacecraft. I said, "Come on you guys, that could be anything." Then they said, "Just wait." All of a sudden a few of the kids standing around the computer gasped for air and jumped back. Suddenly, the spider had started crawling down the spacecraft right above one of the astronauts who was climbing down the Apollo Spacecraft ladder to get onto the Moon surface. The spider was just about to drop onto the astronaut's head and the video ends abruptly. I said jokingly, "That has got to be the greatest discovery of the twenty-first century. You caught a Space Alien live on film, wow!" I assumed the kids had just found some film someone had edited for whatever reason.

One of the kids shouted, "That's nothing, this next one will make your head spin!" When I saw what they had found next I lost my breath for a second. They started playing a semi blurry live footage of the astronauts filming one of the craters on the moon surface next to the Moon Rover. It ran for about 20 seconds and when it was done the kids said, "Did you see it, did you see it?" All I saw was a blurry film of the moon scenery. They said, "Look again, there is a dog running around in the crater." Sure enough, when they played the film again it looked like a dog running around in the background.

At this point I began taking the kids seriously and asked them what other odd things they had discovered. They showed me several pictures of other animals in the moon landing photos including; birds, snakes, cats and even a black dog resting on the Moon rover vehicle. No credible scientist in the world would disagree that a space vacuum with no oxygen surrounds the moon. It would be impossible for any unprotected Earth creature to survive on the surface. The kids later decided to name this first space alien ever caught live on video "Travis Moonius" after Travis my youngest of seven boys.

NASA ID# a15trailer

I was now in shock. It was clear to me NASA had staged at least part of the Apollo Moon Landings and some, if not all, of the twelve astronauts that claimed to have stepped foot on the moon, may be lying. I have always lived by the Hawkins Law, "If one piece of evidence proves without a doubt that something is true or false, all other evidence will ultimately support that conclusion."

Notify the News Media

The kids started shouting, "You need to let everyone know that NASA faked all the Moon landings." One kid added emphatically, "Yes, Mr. Hawkins you need to let everyone know." Others replied, "Call the newspapers and TV stations and let everyone know the moon landings were faked." "Are you crazy?", I said. You guys said it yourselves; you think NASA has killed other whistle blowers that have tried to expose the Apollo moon landing hoax, and, quite frankly, I don't want to be another one of them." "Besides, you've seen how everyone, including your parents and teachers acted when you tried to explain to them why you believe NASA never went to the moon. Even though you have the proof, it would be almost impossible to get anyone to listen or absorb the new material. Besides, NASA must surely have an entire department dedicated to covering up stories like this.

Come to think of it, the American Government has always kept the moon landings somewhat of a mysterious event, even though it was supposedly the most significant scientific event in world history. The news media are certainly censored from broadcasting evidence that is vital to national security. Can you imagine what would happen if word got out that the Americans faked part, if not all of the Apollo Moon Landings?

"You can't let the government keep lying to us," the kids shouted. A young man named Phil argued, "How can we make educated decisions in the future when we are basing them on bogus scientific information put out by the American Government?"

They were right! That last statement was the deciding factor that convinced me to organize the research of the entire Apollo Moon Landing Program and write this book in hopes of exposing the Moon landing hoax for the good of our children's future. Someone needs to be willing to stand up to big government. Otherwise, their deception will continue to impact the world's future in a negative way. Surely, we don't want our future world leaders basing their decisions on a fairytale backed by falsified scientific data.

NASA's deception has already altered our world in many different ways. For instance, many children growing up in the 1970's and 80's fantasized about becoming astronauts, rather than doctors. If they had become doctors, one of them may have already discovered the cure for cancer, but we will never know. As a concerned parent, I feel it is my obligation to devote time to something of such importance.

Recruiting More Help

The discovery of these Earth creatures on the moon started a whole neighborhood scavenger hunt, which began to spread quickly. However I realized that if we were going to convince the world that NASA had faked the Moon landings it would take a great deal more compelling evidence than just a few videos. I told the kids we needed two things to get started. First, we would need an enormous amount of additional evidence including Apollo Mission photos, films, and equipment specifications. Second, I would need a team of Scientists, Engineers, and Geologists to confirm all of the findings. They said they could get everything I needed from the Internet.

Soon they had kids from all around the world joining in the search through the Internet. There were hundreds of kids of all ages working feverishly to find anything they could on the moon landings. Each kid was hoping they would find something that could be used to help prove to the world the moon landings were faked. Rumors of our small group's quest spread like a wild fire, before I knew it, the Internet WhizKids had scientists, engineers, and teachers from all corners of the globe contacting me, offering to help expose the moon-landing hoax. This provided an excellent opportunity to collaborate with some of the most reputable scientific minds from around the world.

Let me tell you, I could hardly believe it myself; some of the things these kids found were amazing. We're talking about hundreds of mislabeled NASA photos and films which clearly indicate the moon landings were faked. There are photos showing levitation harnesses, and wires being used on the alleged moon surface, tears in the astronaut's space suits, and much more. They also found NASA Apollo mission equipment specifications and records, which after examination, indicated NASA was clearly exaggerating their technical capabilities. The WhizKids from Japan even discovered what appears to be the movie script used by NASA for the Apollo 16 Moon Landing mission. Secret messages left behind by former NASA employees working on the Moon Landing project were found. These NASA employees risked their lives leaving these clues in hope that in the future someone would eventually find them. They will be excited to hear we have found most of their clues, and have dedicated a whole chapter in this book to reveal their clues.

The one thing we did not find is based on a rumor started back in the 1970's. Supposedly one of the videos NASA originally released contains two mice running around on the moon's surface while the astronauts were working in front of the Apollo LM spacecraft. The NASA Moon Landing film footage containing dogs, cats and a large spider have already been found, but not the one with the mice. This NASA film is needed to complete the collection for the Moon Landing Bloopers film currently in production. So a reward has been offered for the mice video. You can find the details at www.moonbloopers.com or www.whizkids.tv. Hopefully this special video can be found before NASA officials destroy all of their Moon landing "evidence," in an attempt to cover up their hoax.

The search for the truth began almost two years ago, as our team worked together relentlessly to solve this greatest mystery. The group's efforts are now referred to as the "WhizKids Project". I have written this book, in hopes that all this hard work will not be wasted.

Right now you're probably saying to yourself, "Just one minute, how is this author going to find all this evidence, decipher it, prove it as fake, and then get others with power to help encourage NASA into admitting the moon landings were fake?" Besides, what could be this author's qualifications that happen to make him an expert on NASA space missions? Certainly, the complexity of space missions involves many different aspects of advanced scientific expertise. This individual would have to be able to communicate with other experts in each field. "Yes, if I were just another crackpot making unsubstantiated claims on a subject out of my capability and knowledge, then that would be a problem. So to help reassure you that the results in this book are conclusive, allow me to provide you my background and qualifications before we begin addressing the controversial and never-before-seen evidence my research has uncovered.

I have been involved in many aspects of engineering and science most of my life. During my formative years in the 6th and 7th grades while attending Monroe Elementary School, I was selected for a special U.S. Government program for advanced students. I was sent to a special United States school with a select group of students from other surrounding school districts. The school was referred to as Como Park technical college and was set up as a science laboratory where we focused on applied science. For two years, I spent most of my 6th and 7th grade studying and conducting experiments related to physics, chemistry, biology, geology, including an in-depth review of the principles of electricity, magnetism, radiation, astronomy and space vacuums.

Today, this type of schooling is very common in most large metropolitan cities around the United States. Thanks to these types of school experiments in the 1970's, parents today can send their children to schools that focus on science, art and music in hope of fostering their educational experience.

This early stage special schooling helped me to be considered one of the best electrical auto mechanics in our area at the age of 14, working on just about every type of car around, including maintaining local police department vehicles. At age 16 and 17 I attended college for two years studying industrial mechanics, geometry, trigonometry, calculus, physics, statics, dynamics, thermodynamics, fluid mechanics, material compositions analysis, geology, chemistry, electrical heating, plumbing and air conditioning.

I later attended two more years of college at Northwestern Technical Institute and earned a degree in Computer Science, with a major in robotics and industrial electronics. Since then, I have spent nearly two decades in the field and have devcloped several products, including computer software products while continuing my education. I am currently researching the un-natural acceleration of nuclear energy in Earths Magnetosphere to determine it's potential effects on the inner orbit, the ozone layer and the planet's ecosystem.

Furthermore during the Apollo Moon Landing Mission era my father was an aircraft mechanic for the air force and my grandfather was a retired top ranking aviation inspector for the US air force as well. My grandfather claimed his father was also an aircraft mechanic; however that may not be accurate since no one in the family ever met my great grandfather, it's believed he remained in England while the rest of the family move to America somewhere. Also my grandfather was very old at the time he mentioned this and his recollection of events does not match any type of historical record. For example my grandfather would occasionally start mumbling, "you know laddy, the Wright Brothers were not the first people to invent and fly an airplane. It was your great grand-pappy and his friend william who successfully built and flew the first motorized airplane, almost 10 years before them there Wright Brothers boys.

Obviously grandfather must have been suffering from Alzheimer's disease at the end; nevertheless in his peak years he was considered one of the best aircraft engineers in the United States making aviation a part of our family for generations. This general knowledge combined with my educational background provides me with an in depth knowledge to make a professional evaluation and an accurate scientific assessment of the authenticity of this subject. If the WhizKids could gather enough evidence and had the right amount of specialized help, we could prove conclusively whether or not the moon landings were faked.

As the evidence poured in, it became most exhilarating overseeing the research and writing this book about the American's Moon landing hoax. I still remember the big change in everyone's attitude when it became more and more obvious the moon landings were faked. It was amazing to feel the excitement in the neighborhood as the investigation and our research came to a conclusion. It was the craziest thing around our house; everyone was just clamoring to help out and get involved. The neighbors, their kids, and even the kids' teachers all wanted to pitch in. It was at that point they stopped referring to me as the neighborhood's nutty professor and started calling me all sorts of funny nicknames like "Defender of Modern Science", the "Worlds Greatest Detective", and even Albert Einstein the 2nd (I guess because of my short stature and resembling hairstyle). Occasionally, I would tease some of the teenagers by not revealing my educational background and tell them I was a time-traveler named Micki, just to see if any of them could figure it out.

Okay that's enough about me, what about all of NASA's evidence claiming that they landed man on the Moon back in the 1960's and 70's. Unfortunately, after over two years of hard work and research of NASA's evidence, absolutely NOTHING has been found to indicate the moon landing events were real. All the evidence, after careful scrutiny, supports the theory that the Apollo Moon Landings were the greatest magic tricks ever performed, fooling billions of people worldwide. When the Moon Landing evidence is scrutinized and examined under the microscope, it reveals many examples of conditions that simply cannot exist on the moon. No matter how big NASA's budget, they can't change the laws of physics. The original NASA photos and films show all kinds of things including stage structures, stage lights, underground plumbing, levitation harnesses and much more. There is an abundance of evidence proving virtually everything NASA has told us about the Apollo Moon Landing missions is a lie.

In addition to all the Moon Landing photography being faked, the capabilities of the Apollo Spaceship were also greatly exaggerated. For instance, the Apollo Spacecraft was customized with an escape system to sneak the astronauts off the spacecraft before they were supposedly launched in to outer space. The astronaut's escape took place right in front of everyone's eyes and no one ever noticed. An escape trick even the famous Harry Houdini himself would have been proud of.

Even many of the comments made by the astronauts during the moon landing missions are simply ridiculous, such as seeing space aliens, which points to the fact that they are not genuine. Many questions have been put to NASA about their claims of alleged Moon landing technology, and they have yet to reply with one logical explanation that would pass a modern day scientific evaluation. NASA's policy is to refuse to answer difficult questions about the Apollo moon landing missions.

What about the astronauts? Why are they reluctant to talk about their alleged moon missions in detail? If there is nothing to hide and NASA televised the event live to the rest of the world, why can't the astronauts answer questions? If NASA was telling the truth about landing twelve men on the moon, then they could easily disprove the allegations in this book in five minutes, yet they refuse to do so, why? Instead they ordered almost all evidence related to the moon landings classified until the year 2026. Is NASA simply hiding the truth until those involved have all died of old age, and no one is alive to blame?

Although a great deal of NASA's evidence was quite a challenge to decipher, you'll find as we go through the evidence most of NASA's methods of trickery are easy to understand. All it takes is some common sense and a little basic scientific logic to confirm the Moon Landings were nothing more than an elaborate hoax put on by the American Government. Even in the best of circumstances, the only proof that man went to the moon is the photos, film footage and moon rocks that NASA gave us, and those are riddled with deception.

CHAPTER 2
CONSPIRACY THEORY

The evidence contained in this book could affect the national defense of the United States and should have been classiefed under the Espionage Law, Title 18, U.S.C., Sections 793 and 794, what was NASA thinking?
UN-CLASSIFIED INFORMATION

Chapter 2
The Conspiracy Theory

More than three decades after Neil Armstrong supposedly made his giant leap for mankind, two questions still keep hounding us. Could NASA and the American government mastermind a hoax of this size? If so, how could it have gone undetected for so long? It seems unlikely when you consider the limited amount of evidence NASA has released to the public as proof. The difference between NASA sending men to the moon and staging the landings here on Earth is a trick much like removing one or two playing cards from a deck of cards.

Using a deck with a few cards missing, many card games can be played for years and no one would ever know the difference. The only person that would know is the individual who removed the cards. When it comes to the Moon landing game, that person was NASA and the American Government.

Let's stop and think about this for just one minute. For over thirty years, NASA has been the dealer of the Moon landing game. They have removed all the cards they didn't want the other players to have. This gave them an unfair advantage. Unfortunately, there is almost nothing we can do about it since they have been refusing to let anyone get a good look at their deck of cards. What are we to do? NASA claimed all the cards were there and asked everyone to take their word on it. When anyone asked to see the full deck of cards, or all the evidence related to the moon landings, NASA refused. Isn't that ironic.

Should we be trusting NASA? Are they playing with a full deck of cards. I don't think so. After looking at all the cards NASA has dealt to the public so far, I can tell you this. They have removed over half the cards from the deck and stored them for safekeeping.

Sadly, the only way to know which cards NASA has removed is to go through the cards they gave us and see which ones are missing. That would seem like an impossible task, but if that's the way NASA wants to play the game, then it's about time we call their bluff and count the cards. Wouldn't you agree?

This book takes you through the process of checking every card NASA has dealt to us. Let's start with the Ace of Spades, which represents the top-ranking person who was in charge of the Apollo Moon Landing Project. We need to know who he was, where he came from, where he lived, his educational background, and his work experience.

It took a long time to find NASA's Ace of Spades and the result was shocking. After discovering who NASA's top-ranking official for the Apollo Moon Landing project was, it's hard to imagine why no one ever seriously questioned the authenticity of the American Moon Landings. If the American government was able to convince us that in the year 1969 they had the secret technology to take us to the Moon, then certainly faking the event would have been a piece of cake in comparison.

The number one man in charge of NASA's Apollo Mission team of engineers and scientists was the amazing Werhner von Braun. It is almost impossible to imagine that he was the top-ranking scientist in charge of the Apollo Moon Landing project.

Why did NASA consider Werhner von Braun to be the most qualified person in the world to oversee the Apollo program from start to finish? The American Government obviously believed Werhner von Braun had the experience and special talents to make the Moon Landing project a success.

Head of NASA's Apollo Project
Werhner Von Braun
NASA ID# ap11-KSC-69p-632

After accepting the challenge, Werhner von Braun went to work immediately, first assembling a special team of talented people to work together on the Apollo Space project. This special team was made up of many professionals including some of the world's best scientists, photographers, artists and filmmakers. But how was Werhner von Braun going to get the job done at the time NASA only had about 300 rocket scientists, compared to the Soviet Union, who had over 30,000!

How could Werhner von Braun perform such a task with so few people? Was he some kind of super human? The Soviet Union was unable to send humans to the moon, and they had 100 times the workforce dedicated to space travel. What was NASA thinking? Was the American Space program that superior to the Soviet Union's back then?

Regardless of how understaffed NASA was at the time, it's clear that NASA did pick the right person for the job. Werhner von Braun and his team, codenamed "The Monday Project", by some miracle appeared to have completed the task! We have all seen the Apollo Astronauts blasting off in their rocket, floating in space, and walking on the moon. None of that would have been possible without Werhner von Braun and his team of experts.

Nevertheless, one has to wonder how von Braun and his team really accomplished such a feat. Werhner von Braun's team looked very suspicious, and so did the entire NASA program. Did NASA put the right man in charge of the Apollo Moon Landing project? For instance, why did Werhner von Braun believe it would require almost as many artists, photographers, and filmmakers as scientists on his team to carry out a manned mission to the moon?

Was NASA planning to send a team of photographers with the astronauts to the moon to take pictures of the astronauts working? Did the government send photographers to the war fields? No, NASA scientists were not planning to send photographers to the moon; they were planning to make a movie! That's what made Werhner von Braun the right person to oversee the entire American Apollo Moon Landing Project.

NASA Plans to Make Moon Landing Movie
Werhner von Braun was the World's most famous space movie producer.

Werhner Von Braun working at Walt Disney on space movie July 1954 Photo from NASA

Before being hired by NASA, Werhner von Braun was a world famous movie producer and actor working for Walt Disney Studios. He was considered the world's greatest space movie technical genius at the time. He produced "Man in Space", "Man and the Moon" and "Mars and Beyond". These were three of the most popular science fiction space travel movies from 1955 through 1965. His space movies were so realistic at the time that many people would not be able to tell the difference. He was known to have the space adventure imagination to the point that he had most people convinced of his prediction that a baby would be born on the Moon by the year 2000.

Werhner von Braun's first television series was "Man in Space" released March 9, 1955. His first movie, "Man on the Moon" was also released in 1955. It was a movie about a trip to the moon in a rocket ship. Need we go on? "Into it went the thinking of the best scientific minds working on space projects today, making the picture more fact than fantasy.", said Disney producer Ward Kimball. He also claimed that when Werhner von Braun's movie first aired, President Eisenhower immediately called Walt Disney to compliment him on the show. He requested a copy so that he could show it to top space-related officials in the Pentagon. It was so realistic that the representative of the Russian space delegation L. Sedov requested a copy on September 24, 1955 from the President of the International Astronautical Federation.

"Mars and Beyond" was Werhner von Braun's second major space travel movie. It was released on December 4, 1957. It showed a thirteen-month manned mission to the big red planet Mars.

Evidence suggests that once NASA was certain there was no possibility of a manned trip to the Moon, Werhner von Braun invited Walt Disney, his brother Roy, and other Disney executives to the Marshall Space Center. It is believed that part of the filming was taking place there for the upcoming Apollo Moon Landing missions at the time. Was Werhner von Braun trying to convince Walt Disney Studios to help produce the Apollo Moon Landing movies for the good of the country?

It's impossible to be sure, but after that visit Walt Disney was quoted in The Huntsville Times as saying, "If I can help through my TV shows . . . To wake people up to the fact that we've got to keep exploring, I'll do it." It is widely believed that Walt Disney was referring to helping NASA make the Apollo Moon Landing movies. Uncovering Werhner von Braun's background history was very valuable, and helped solidify the theory that the Moon landings were a hoax put on by the American Government.

Werner Von Braun: NASA appointed Leader for the Apollo moon landing project

Was NASA capable of such an elaborate Hoax?

Even though NASA had the right people in place to stage the moon landings here on Earth, could they have performed a hoax of such magnitude without being detected? Not even one leak by someone. The answer is YES, of course. Absolutely! Remember, we're talking about the American Government. It would have been easy considering the way the American Government kept many of its most top-secret projects independent of one another.

Remember, in many cases, the American Government constructed massive military facilities called "Secret Cities" to contain the information. The American Government used this practice as far back as the development of the hydrogen and nuclear bombs. More than 10,000 people were sworn to secrecy while working on the bomb projects. They all worked in large complexes and once they left the "Secret City", they were not allowed to speak a word about what they were working on. It is believed that the CIA spied on the workers and if someone was caught talking about the projects, they could be immediately imprisoned. Furthermore, they would never be seen again. The hydrogen and nuclear bomb programs were so secret that many of the people working on the project claimed they did not even know what the goal was until after the bombing of Japan. Only then they knew.

Later, during the Apollo Moon Landing years, NASA used similar strategies for spy satellites, space shuttles, Skylabs, and Stalk Boomers for the Apollo Moon Landing Missions. No question about it, NASA was a master at keeping its projects isolated from one another. Our research indicated that with the manpower of less than 200 workers, NASA produced the moon landing movies in secret. They did this with training mission simulators.

With only a few hundred people involved with the hoax portion of the American Moon Landings, it was very easy for NASA to keep it covered up and undetected for so long. Surely there were many ways for NASA to silence the limited number of people involved in the actual Moon Landing Hoax. Many believed NASA refused to pay these employees their retirement pensions. They could threaten to imprison them and in some instances even threatened to kill them. What happened to many of the original Apollo astronauts who spoke out about the problem they experienced while testing NASA's space equipment? They died in suspicious accidents and fatal mishaps. Many people, including some family members, believe these deaths were arranged by NASA to conceal the truth of the Moon Landing Hoax.

At the time I wrote this book, most of the evidence used to expose the Moon Landing hoax had been obtained from NASA's own archives and was still available to the public. Unfortunately, shortly after this book is published, NASA will probably have destroyed much of the incriminating evidence and will most likely have killed me in an attempt to conceal the truth. In anticipation of these two events taking place, I've stored copies of all of my backup research documentation in a secret location. Nobody knows the whereabouts.

I have encrypted into this book a secret algorithm code revealing the location of the materials proving that the Moon Landings were a hoax. Along with these materials there is a very large sum of money to assist whoever finds the information. This money will help keep the fight for truth alive so that someday the American government will have to admit its wrongdoing, correct the history books, and put an end to this scientific outrage. A lie as big as claiming to go to the moon should not live forever. To help ensure that everyone is looking in the right direction for the treasure, here is the first clue. "Wise are those efforts relying on the guidance of Els to solve this puzzle. You will be led to one of earth's first stellar travelers, a real bear, strongest of all creatures living on the planet earth. For she is the one who holds the treasure you are looking for."

Now, realize it is not going to be easy to get the American government to admit to it's wrongdoing. NASA is already refusing to answer questions about its alleged manned missions to the moon using basic scientific principles. Unfortunately, NASA's policy regarding any allegations of fraud is to respond by accusing such individuals of being conspirators. Of course no rebuttal of the allegations are allowed. Obviously if NASA had such futuristic technology they could easily make all of the people claiming the Moon landings were fake look like fools. This makes you wonder why NASA refuses to respond to any of these so-called crazy hoax conspirators.

Are All The Hoax Conspirators Just A Bunch of Crazy People

Could all of these other Moon hoax landing conspirators be just a bunch of crazy people, like NASA wants us to believe? After researching their backgrounds, it was remarkable to find that they all have one thing in common. These so-called crazy Moon hoax conspirators are some of the smartest scientists, engineers, and astronauts on the planet. Our research shows that many scientists and astrophysicists of today are beginning to speak out against NASA's claims of landing man on the moon, and backing it with scientific proof. Even world famous historian A.J.P. Taylor referred to the moon landings as "the biggest non-event of his lifetime."

As the WhizKids network of students, scientists and engineers grew they told us that in many of their countries they are taught that the Moon landings were scientifically impossible. The leaders of their countries claim the Americans faked the whole thing and then falsified scientific evidence. Japan, regarded as one of the most technologically advanced countries on the planet claims the Moon landings are a big joke. In that country you can stop almost any seven-year-old on the street and be told that the moon landings were faked. If you ask these kids to prove it, they will tell you to look at the ridiculous pictures on NASA's websites. From what I have learned about the Japanese culture. They are known for taking pride in teaching their children right from wrong at an early age. If scientific evidence supports the possibility of landing men on the moon, then why would the Japanese say such a thing, this contradicts every fiber of their culture's ideology.

It is clear there is a great division on the subject as more and more countries are beginning to speak out. They are claiming that the Americans produced a series of TV shows filled with outdated space theories, numerous inaccuracies and unsubstantiated conclusions. Countries such as Japan, Cuba and many others now refuse to support any claims that man landed on the Moon. Many believe it is a disgrace to the American people, the astronauts and the brilliant engineers who worked and are still working today to achieve so many of mankind's greatest technological achievements, like the Space Shuttle missions and the Mars Rover missions to name only a few.

With so much controversy over NASA's Apollo Moon landings, one can only wonder if the Untied States Government was telling the truth. Certainly, during the '60s and '70s the U.S. Government was guilty of many Cold War cover-ups. Officials admitted to using their own people as guinea pigs for electroshock, drugs, hypnosis, sensory deprivation and other types of trauma. The CIA conducted Mind Control experiments on children attempting to create a Manchurian Candidate. Thousands of innocent American children are still silently suffering today as adults from the effects of these experiments.

What has the American's Government done in the past?

American doctors were paid to take part in these and other experiments. They did research with Chemical and Biological Warfare while working under government contracts in the U.S. and Canada. You can read the proof in more than 18,000 pages of declassified documents from Bluebird, Artichoke and the MKULTRA Projects Documentation as well as Testimony from Survivors of these Cold War Tests.

Recently I discovered that several members in my family, including myself, are survivors of cold war testing when the government conducted hazardous chemicals tests during the late 1960's and early 1970's. Supposedly my classmates and I were continuously sprayed over the years with hazardous chemicals by the American Government to determine their short and long-term effects. The survivors and I still remember these tests very clearly and how they were conducted. During recess on the school playground, crop dusting planes would pass overhead and spray us with hazardous chemicals. Since we were all kids we didn't realize that something was wrong with a crop duster flying low in a major metropolitan city, with no farm within a forty-mile radius. We were all so fascinated with them we even referred to them as "Skywriters". We would watch to see if the person in the plane was trying to spell something with the clouds of smoke spraying out behind the plane.

When you think about it, spraying school children with bioterrorism chemicals to test the effects should not be any big surprise, since at the time the American Government was doing things far worse than that. For example, Americans have been exploding nuclear weapons in outer space since the late 1950's and it is estimated that more than 2,000 nuclear bombs have been detonated out there. Does that sound unimaginable? Here is some proof, the fallout from the weapon tests in outer space has actually created two large holes in earth's protective ozone layer.

Admittedly, for the American Government to conceive a hoax as big as faking a moon landing, and for this to go undiscovered for so long, seems impossible. It is really difficult to imagine that the American Government would go through so much trouble and expense. To cover up the hoax, they would have to falsify history books and encyclopedias while teaching students false science for decades. Honestly, would the American Government ever do such a crazy thing as deliberately falsifying technological abilities to fool other countries into believing America was superior? Of course, the American government does it all the time and has even admitted to doing it.

For instance during World War II the American Army used fake look alike tanks that were filled up with air on the battle field to fool the other side into thinking they had a much larger arsenal of tanks than they really did.

Furthermore in 1994, the American government's watchdog agency, the General Accounting Office reported, "The Star Wars Missile Defense System rigged tests to make it seem more advanced than it really was. The aim was to fool the Soviet Union about the United States' strategic capability during the Cold War."

As ridiculous as it sounds, after years of research a great deal of solid evidence supports the Moon Landing Hoax Theory, and shows that the United States Government is continuing the massive task of manipulating information to keep the hoax going. This includes concealing most of the alleged moon landing evidence from the public, deliberately altering the history books with inaccurate scientific principles, and writing articles discrediting people who spoke out against the government's claims of landing men on the moon.

Hopefully, publishing this book will encourage the honest people working at NASA to openly discuss their claims and provide the world's scientific community the opportunity to conduct independent observations, and then make a conclusion that validates or modifies their hypothesis. Certainly, after all these years, NASA must no longer be allowed to put its claims through the rigorous testing necessary for validation. It is NASA's outright refusal to declassify information until 2026 that is creating serious damage to its credibility.

Why has man not returned to the Moon?

When our group of truth seekers began this adventure, the most common question everyone asked was, "If humans were so successful at landing on the Moon six times during the 1960's and 1970's, why haven't they been able to go back?"

As an experiment, I suggested several people call NASA and ask different people working there that very question. The responses we got back from NASA employees were very surprising and seemed to support the theory that NASA faked the Moon landings here on Earth. We received answers like, "I don't know", "It's too dangerous to go to the Moon", "The space shuttles go to the Moon all the time" and "We already have a space station on the Moon". The most common answer was, "We can not afford to go back to the Moon, it would be too expensive."

None of NASA's answers make sense! One can only wonder if some of these NASA employees are trying to give us a clue that sending a man to the Moon is not possible. If NASA's Moon landings were genuine, then going back should be far easier and less expensive today than it was back in the 60's and 70's.

NASA Still Lacks The "Technical Ability" To Send Men To The Moon

The answer to why man has not returned to the Moon is the fact that they were never there in the first place. Even today NASA still lacks the technical ability to go to the Moon. There have been many new scientific discoveries made since the Alleged Apollo Moon Landings that eliminate any possibility of events taking place in the 1960's and 70's. The biggest obstacle is the high levels of extremely deadly radiation recently discovered beyond Earth's immediate inner orbit. These radiation levels are so powerful they would kill any astronaut attempting to travel to the moon, utilizing today's primitive technology.

Many people don't realize that a Magnetic Force Field surrounds the earth and protects everyone from deadly space radiation. This protective force field is called the Magnetosphere and sometimes called the Van Allen Radiation Belts after James Van Allen who discovered the radiation belts which make up a part of it. Earth's Magnetosphere is a lot like a force field of a Star Trek Spaceship seen in the movies with its shields up at all times. It is located 650 kilometers (400 miles) above Earth and extends an additional 65,000 Kilometers (40,000 miles) out towards deep space. To get to the moon, astronauts must travel through Earth's nuclear radioactive force field because the moon is approximately 200,000 miles beyond the force field.

Besides, if getting through the Magnetosphere is not enough, beyond that is deep space, where the deadliest radiation is found. How this type of radiation could affect humans has been studied in space for years and confirms that any attempt by humans to travel through this radiation field and to the moon with such primitive technology would mean instant death for the astronauts.

NASA scientists obviously know space travel to the moon is impossible with any type technology known to humankind. They have been secretly testing the deadly radiation in outer space for many years now with anatomical model of the human torso and head called "Fred" which confirm the astronauts would be killed instantly once they attempted to leave earth's inner orbit.

This Phantom Torso "Fred' contains hundreds of radiation monitoring devices which measure the high-energy particles that would have passed through the human and would disrupt way cells function. The deadly effects of these radiation belts and other forms of deep space radiation is discussed in greater detail later in this book and serves as a good reference

Meet "Fred" NASA's Phantom Torso, who's test result confirm humans can not travel to and from the moon!

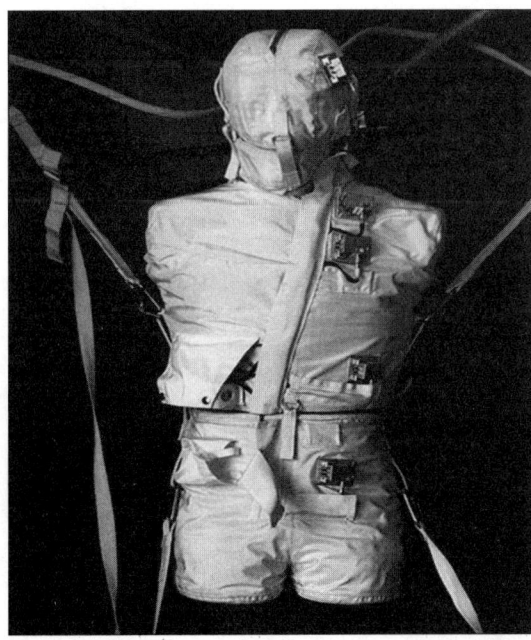
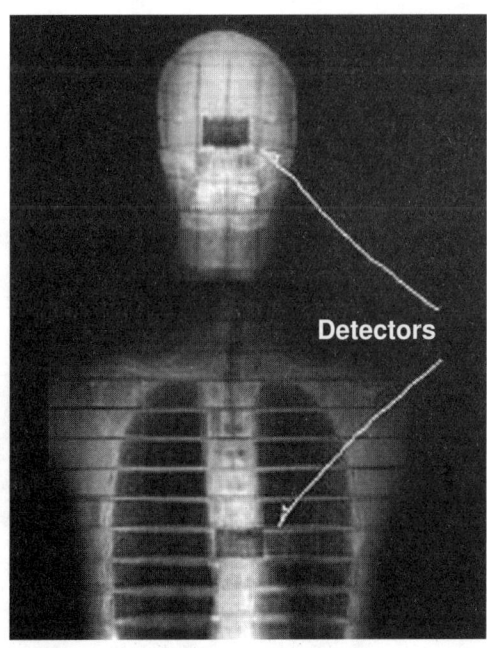

CHAPTER 3

Suspicious Deaths

Chapter 3
Suspicious Deaths

Secret government cover-ups and conspiracy theories involving the Central Intelligence Agency (CIA) including suspected CIA murders are very common in the United States. From the murder of JFK to Clinton's White Water Conspiracy Theory people have been claiming that the CIA has killed certain members involved. The deaths related to the alleged Apollo Moon landing project are no exception and there are more of them because it was a much bigger project.

Usually the suspected CIA murders are just accidental deaths occurring at coincidental times, which just happen to make good stories. As a result the news media prints the murder conspiracy story and soon the public starts to believe it. The story gets bigger and bigger each time it's told, and eventually it becomes a legend. At that point most people cannot tell if the story is true or not, except for the immediate family. All the time the story has been getting bigger and bigger, the immediate family has been saying, "It was an accident" and no one ever believed them.

The most disturbing thing about the suspected government murders related to the Moon landing hoax is how completely opposite the family members reacted. Some of the victim's immediate family members, including widows and children, are claiming the Government killed their husbands and fathers.

Once more, almost no one believes the immediate family because NASA and the news media have convinced people the claims made by these family members are ridiculous. We may never know why people have such a hard time believing something when it comes right from the people that knew the victims the best. To help solve this mystery each suspected murder is going to be examined for any suspicious foul play involved, starting with Thomas Baron and the mysterious death of his family.

Thomas Baron and his Family's Mysterious Death

Thomas Baron was working on a report that dealt with critical issues related to the development of Apollo spacecraft hardware. He became very outspoken after the tragic fire that killed the Apollo 1 crew, and you will see why later in this book. He had already been known to the press as a whistle-blower and a critic. It is widely believed the death of this NASA worker was a murder to silence him about his 500-page report. Some believe NASA then took the 500-page report from Congress who at the time was in possession of it, and altered the contents to cover up the truth. He and his family died when a train struck their car at a crossing. We will never know if it was an accident or if their car was trapped on the railroad tracks by a car in front and in back of them to make it appear to be a suicide. Although, the idea of a husband and wife choosing to park in front of a speeding train to kill themselves seems out of the ordinary.

Suspicious Apollo Astronauts Deaths

Almost Half Of The Apollo Astronauts That Tested The Equipment Died In Mysterious Plane And Automobile Accidents.

Imagine the odds of almost half of the 30 plus astronauts who trained for the Apollo mission, dying in mysterious plane and automobile accidents! Most of these accidents (murders) happened within a two-year time frame of one another. Was NASA afraid they would have eventually leaked information about how the equipment was incapable of taking them to the moon? Because the deaths were all of the same type and during the same time period, many believe NASA and the CIA killed these people to silence the truth, including members of the astronauts immediate family.

Ted Freeman (died in a mysterious plane crash - October 31, 1964- believed to be the only accident)
Gus Grissom (suspicious spacecraft fire - January 27, 1967)
Ed White (suspicious spacecraft fire -January 27, 1967)
Roger Chaffee (suspicious spacecraft fire -January 27, 1967)

Suspicious Deaths

Elliot See (died in a mysterious plane crash - February 28, 1966 T-38 jet crash)
Charlie Bassett (died in a mysterious plane crash - February, 28 1966 T-38 jet crash)
Ed Givens (died in a car accident - June 6, 1967.),
Robert Lawrence (died in a mysterious plane crash- August 12, 1967)
C. C. Williams (died in a mysterious plane crash- October 5, 1967)
Mike Adams (died in a mysterious plane crash- November 15,1967 Plane explodes over California desert)

Charles "Pete" Conrad, Apollo 12 Astronaut, died in a very coincidental motorcycle accident. Many believe he was going public about the fake Moon landings on the 30th anniversary back in July 1999. Coincidentally he was killed in a suspicious motorcycle accident one week before the 30th anniversary.

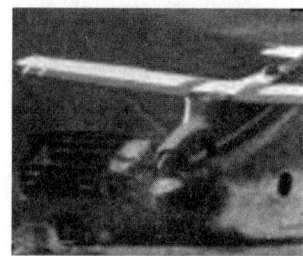

Alan Shepherd and his wife died suddenly of cancer just a few weeks apart. Some believe NASA killed Alan Shepherd because he was ready to talk about the hoax. After learning of this possible murder, I immediately thought about the likeness to Jack Ruby, who was connected to the President Kennedy murder. Jack Ruby told his family members he was injected with cancer cells by CIA operatives.

Gus Grissom, Ed White, And Roger Chaffee Die in a Suspicious Apollo 1 Fire

Gus Grissom, Ed White, and Roger Chaffee died in a suspicious Apollo 1 fire. It is widely speculated by many hoax theorists that on January 27, 1967 the Apollo 1 fire that killed these three astronauts was a deliberate act by NASA to silence Gus Grissom. Grissom himself even predicted that he would be killed over the Apollo program just three days before his mysterious death.

Betty Grissom, the wife of Gus Grissom, believes NASA murdered her husband. She has long said NASA covered up what really happened in the launch pad fire that killed her husband and fellow astronauts, Ed White, and Roger Chaffee during an Apollo 1 dress rehearsal on January 27, 1967.

She claims investigative records showed the fire was no accident and refers to the event as murder. Gus was often very outspoken about the problems he was experiencing with the Apollo Moon landing equipment. According to the Phoenix Educator News Paper, shortly before his death, Gus Grissom had taken a large lemon and hung it around the space capsule as the press looked on. Gus then supposedly claimed the project could never be accomplished on time. The Associated Press reported, "Pretty Slim" was the way Grissom put the chances of meeting the mission requirement.

Clearly having someone as outspoken as Gus around would be a problem for NASA if the moon landings were going to be a hoax. Did NASA officials believe that Gus Grissom was about to go to the media and expose the hoax and then silence him?

Starting a fire in the cockpit of the Apollo 1 spacecraft was the solution to NASA's problems they were experiencing with the space program at the time. The Apollo 1 would likely be destroyed and guarantee that Grissom, and the other two astronauts would be killed. This would also buy NASA desperately needed time for the Apollo Project. The first flight to the moon was scheduled to take place just one month later on February 21, 1967 and they were nowhere near ready.

Suspicious Deaths

Scott Grissom, son of Gus Grissom, claims NASA murdered his father. Scott Grissom reported in the February 16, 1999 issue of the Star Newspaper that he felt the motive for his father's murder might have been related to his father sinking the American spacecraft, Liberty 7, in the Atlantic in 1961. "My father's death was no accident. He was murdered", Grissom said when he recently was granted access to the charred capsule and discovered a "fabricated" metal plate located behind a control panel switch. The switch controlled the capsules electrical power source from an outside source to the ship's batteries. Grissom argues the placement of the metal plate was an act of sabotage. When one of the astronauts toggled the switch to transfer power to the ship's batteries, a spark was created that ignited a fireball.

A private detective hired to investigate the incident believes NASA murdered Gus and the other two astronauts. Clark MacDonald, a McDonnell-Douglas engineer hired by NASA to investigate the fire, offered corroborating evidence. Breaking more than three decades of silence, MacDonald says he determined that an electrical shortage caused by the changeover to battery power had sparked the fire. He says that NASA destroyed his report and interview tapes in an effort to stem public criticism of the space program. "I have agonized for 31 years about revealing the truth, but I didn't want to hurt NASA's image or cause trouble", MacDonald told the paper. "I can't let one more day go by without the truth being known."

Gus Grissom poses in front of his Liberty Bell 7 capsule before the July 1961 mission. NASA image

Gus's widow, Betty, told the Star Newspaper that she agrees with her son's claim that her husband had been murdered. "I believe Scott has found the key piece of evidence to prove NASA knew all along what really happened, but covered it up to protect funding for the race to the moon."

The Whiz Kids asked that we dedicate this book to these service commemorated astronauts; Gus Grissom, Roger Chaffee, and Ed White, who were killed on January 27, 1967, as the blaze engulfed the Apollo 1 capsule. Also Betty and Scott Grissom for their dedication to find the truth related to the NASA program.

More people have spoken out about the cover up including Bill Kaysing a nonbeliever who got in touch with his friend Paul Jacobs, a private investigator from San Francisco. He was asked to help him with his Apollo differences investigations. Mr. Jacobs agreed to go and see the head of the U.S. Department of Geology in Washington. As he was traveling there the following week after his discussion with Mr. Kaysing he asked the geologist if he has examined the Moon rocks and if they really came from the Moon. The geologist just laughed. Paul flew back from Washington and told Kaysing that the people in high offices of the American Government knew of the cover-up. Paul Jacobs and his wife both died of a mysterious cancer within 90 days!

Bill Kaysing claims another friend of his Lee Gelvani, said he almost convinced informant James Irwin, an Apollo Astronaut to confess about the cover-up. Irwin was planning to tell Kaysing about his finding, however he died 3 days later!

Suspicious Deaths

The Murder of President John F. Kennedy

Although there are many theories about the murder of JFK, there is evidence supporting one more and hopefully the last possibility. After a thorough investigation of the Apollo Project, new evidence strongly suggests President John F. Kennedy's murder was related to the Apollo Moon landing hoax. This was totally unexpected, but so many things relating to the Kennedy murder fall right into place.

The JFK MURDER

Here is the new JFK MURDER THEORY that was uncovered during the investigation of the Apollo Moon landing hoax. Please note that this is just a theory for now and all allegations have yet to be confirmed or denied by government officials. However this theory is the only one that makes any sense and is supported with some very convincing circumstantial evidence.

On February 14, 1961 James Webb accepted the position of administrator for NASA. Just a few months after he accepts the position, he starts insisting that President Kennedy push congress for the funding to send men to the Moon. Eventually, Webb talks the president into addressing the Nation asking them to put their hard earned tax dollars behind NASA's lofty goal. This was President Kennedy's historical May 25, 1961 challenge to the nation to send astronauts to the Moon and return them safely to Earth by the end of the decade. Just a year and a half later, in November 1963, James T. Webb insisted President Kennedy make a visit to NASA's Launch Operations Complex to discuss an urgent top-secret matter that needs his immediate attention. Below is an actual picture of President Kennedy and NASA's Administrator James T. Webb November 1963 meeting at the NASA facility. This is one of the last pictures taken of President Kennedy before his death.

President John F. Kennedy and NASA Administrator James T. Webb at the Launch Operations Complex (later renamed the Kennedy Space Center) during a tour of NASA facilities on **November, 1963.** (NASA Photo, available on NASA Image Exchange:

This would be President Kennedy's last visit to NASA. During this closed doors meeting with NASA's top ranking officials John was given the bad news that a manned mission to the Moon was impossible. Webb says, "You see Mr. President, we accidentally made one big mistake. We had scientists working on the Van Allen radiation problem I was telling you about. They tried to clear a path through the radiation belts with a few megaton nuclear bombs and unfortunately the results were catastrophic. There is no longer any chance of sending men to the Moon".

"You what?" said the president. Webb quickly replies, "Don't worry Mr. President we have come up with a way to fake the Moon landing and fool everyone into believing we were there. I found the perfect guy to pull it off, he makes astronaut Moon landing movies for Walt Disney Studios. He is the best in the world and is ready to get started just as soon as you give the go ahead. Let me introduce you to Mr. Werhner Von Braun."

3-5

Suspicious Deaths

As you would expect, President Kennedy became very disgusted with the idea of temporarily fooling the American people and the Soviets into thinking the U.S. space technological abilities were far superior. He understood that any hoax was doomed to eventual failure, especially one this elaborate. Someday someone would eventually figure it out. The devastating effect the exposure of a hoax would have on the reputation of the United States would be many times more severe than simply telling everyone a trip to the Moon was not possible at this time. One can almost hear Kennedy saying, "No I will tell the American people landing a man on the Moon is not possible."

Kennedy and Von Braun

Then the question came was brought up by one of President Kennedy's most trusted cabinet members: "What should we do with the 20 billion dollars already being collected to be set aside for the Moon Landing Project? Mr. President, we surely don't want to give it back to the American people." Then James Webb NASA's Administrator, who had been hounding President Kennedy about this stupid Moon landing scheme for the last two years, interjects saying, "think about it, staging a Moon landing hoax would cost less than a billion dollars. The remaining 19 billion dollars could be used to pay for so many other worthwhile NASA space projects like the space shuttle and the Mars landing rovers." Once again President Kennedy's most trusted cabinet member jumps into the conversation and says, "There would also be plenty left over to fill our entire family's pockets full of money. Just imagine there would be enough money left over to buy upstate New York if you wanted."

As expected President Kennedy responds in a very stern voice: "No, we are going to return the billions of tax dollars back to its rightful owners, the hardworking American people". JFK then said, "Now I am going to return to Washington and give you until Saturday to clean this mess up. Then I am going to schedule a special press conference to make the announcement and return the taxpayers money. I want you people to get all of your information together, and be prepared to answer questions from the media. And remember, I want you to tell the American people, the whole truth, and nothing but the truth. Am I making myself clear?" Everyone in the room except for Webb responded with a rowdy cheer, "Yes, Mr. President we hear you loud and clear!" The presidents orders, of course never took place.

Unfortunately President Kennedy did not realize the level of risk NASA would be willing to take. The day before NASA was to come clean about the Moon landing being impossible, a few corrupt NASA officials and high ranking US government officials had President Kennedy murdered during the now famous November 22, 1963 Dallas parade. Of course we know how the story ended: the 20 billion dollars was never returned to the American people. Here is another example of how "Money can be the root of all evil." This would also explain allegations that several government officials at the time started up companies that charged the government ridiculous sums of money for various items like $300.00 for $1.50 screwdrivers and $150.00 for toothbrushes.

Money was not the only driving force to keep the hopes of the Americans alive about sending men to the Moon. The American Moon Landing Hoax did more good than harm; it also played a valuable part in world history. Back in the 1960's, space exploration was clearly in its infancy and general knowledge of space at the time was limited. The Soviet Union was advancing so far ahead of the United States in space technology it was scaring the American people. Somewhat, prompted by a dent in national pride, President Kennedy in May 1961 announced the goal of landing a man on the moon and returning him safely to Earth before the end of the decade.

At the same time, America's military was afraid that other countries were drifting toward communism. They were impressed with the Soviet space success rate and believed that the Soviet Union would eventually be victorious as the world leader. American military leaders would do almost anything to stop this movement toward communism, no matter how bizarre the plans. They desperately needed to create a sense of urgency mixed with national pride to stop the spread of communism. If they could not go to the Moon, then faking it was the next best thing, even if it meant killing one unwilling American President along the way. After all, it was for the good of the country.

Without a doubt, the assassination of JFK and later Bob Kennedy reinforced the CIA murder theories. If the dysfunctional and troubled Oswald was working alone when he killed John Kennedy, then who killed Bobby Kennedy later? What are the odds of two Kennedy brothers being killed in the same manner by two dysfunctional murderers? A more likely reason for his murder was that Bobby stated that one of his primary goals was to open a congressional investigation into the death of his brother if he became president. It makes sense that the people that were planning to fake the Moon landings and keep the American taxpayers money would also have Bobby Kennedy killed.

Modern forensic evidence suggests the possibility that two different people shot JFK from different directions. Presumably one of the gunmen hit the President in the neck and the other in the head. If this is true then the government's claim that Lee Harvey Oswald shot JFK makes no sense, if he was a dysfunctional loner. There would have had to have been two dysfunctional and troubled people that came up with the same crazy idea to kill the President on the same day at the same place and firing at the exact same time. That would be impossible! It is more likely that Oswald was simply in the wrong place at the wrong time and took the blame. While the CIA assassination gunmen simply walked away scene after completing his assignment.

The recent discovery suggesting possibly two different gunmen involved in the shooting does support the CIA assassination theory. What a better place than a parade with thousands of potential people to blame as the shooter. Two well trained assassination gunmen could have easily waited for Kennedy's head to be lined up with a predetermined object along the parade path and they simultaneously took their shots.

The true story may never be known since the United States government most likely has classified the most sensitive documents indefinitely. The U.S. Government takes pride in how efficient they are in keeping information secret from the public. According to official estimates, in 1994 the government took 6.3 million classification actions, creating an estimated 19 million pages of classified documents. More than 32,000 government workers are employed full-time to determine what should be secret, what level of secrecy the material should have and whether the documents should be classified. Currently there are hundreds of millions of pages of secret documents held by the government.

Message reads "We killed KENNEDY....
NASA photo ID # NASA ID # as15-82-11160

However, with so many people involved there is always room for error, and this original unedited photo obtained from NASA is a perfect example. Someone has written in the moon dust a coded message that reads, "We killed KENNEDY. Unfortunately, on printed material this writing message is almost impossible to see.

The good news is the digital image is available for view via the Internet. At the time this book was written this photo could be accessed from NASA's website. Although, by the time you are reading this book, NASA has removed this photo and it is no longer available from them. But don't worry you should be able to still see a copy of this rare original NASA photo at www.moonbloopers.com or www.whizkids.tv.

The Unknown Witness

There is one more very special person the Whizkids requested we dedicate this book to. The last person is the most important person of all; "The Unknown U.S. High School teacher".

This story was sent to us from a gentleman named Mark. He claims he had a female high school teacher in the 70's who believed the moon landings were faked. She was telling his class the details. After school one of the students went home told his parents everything. A day later the teacher disappeared. The school officials claimed she was fired and she was never to be seen again. When a few of the local towns people went to this teacher's home too check up on her, the house was completely empty. Mark claimed to have attended a High School near the Kennedy Space Center where a large part of the Moon landings are believed to have been faked. Is this a coincidence? I don't think so, and hearing stories like this helped us realize one of the major reasons why the American Moon hoax has gone undetected for so long.

NASA Censors All Moon Landing Evidence Released to Public

It was mentioned earlier that NASA regulated all of the independent press coverage for the six manned trips to the Moon. NASA and the American Government strictly controls all of the pictures, videos, and audio shown to the public. Why does NASA act so secretively? They had no firsthand witnesses to any Moon landing other than the astronauts that claimed to have been there. Yet strangely enough, they have refused to provide any substantial information publicly about the events, and refused to give any televised interview detailing their missions since their alleged return from the moon surface decades ago. This seems to be an elaborate scheme on NASA's part to conceal the truth. If the moon landings were faked, there is no way any of the astronauts could be expected to answer any of the more difficult questions related to the mission. Most of the information NASA had fabricated about their missions was based on disproved scientific principles. One example was that NASA claimed the Apollo 12 astronauts had salvaged parts off an earlier spaceship that was left behind (the Surveyor 7) during their mission. Supposedly the parts were easily interchangeable between spacecrafts. NASA had the Astronauts get into their Moon rover, drive over to the other spacecraft, get their tools out and do the work with no problem right there on the Moon surface.

Even more ridiculous was how often the Apollo Astronauts claimed to see building structures and UFO's during their missions. Apollo 11 Astronauts Neil Armstrong and Edwin "Buzz" Aldrin spoke as if they were seeing UFO's shortly after the historic landing on the Moon in Apollo 11 on July 21, 1969. In the book, "Celestial Raise", by Richard Watson, and ASSK, 1987, pages 147-148 records the conversation, which was picked up by hundreds of ham radio operators in the USA.

With today's technological breakthroughs, NASA's claims of the Apollo Moon Landing missions would seem ridiculous. Which explains why NASA refuses to let the astronauts be questioned independently and uncensored, There hoax would be exposed in the first five minutes, no doubt about it. Can you imagine the Apollo Astronauts trying to answer questions like, "what did the UFO's look like that you claimed to have seen? Why did you say you could not see any stars in outer space? Why didn't you bring flashlights to a place as dark as the Moon? Why aren't you deaf after riding on the LM rocket engine or burnt to a crisp? Why were you wearing elevated boots with a 2-inch false bottom during training? Was that so the artificial moon dust could be pasted under you? Why do we see many Moon landing pictures with Earth's creatures in them such as rabbits, snakes, spiders, mice, lizards and birds?"

NASA's evidence of proof is so riddled with errors that anyone can prove the Moon Landings were faked by simply focusing their attention on the authenticity of NASA's own evidence. Almost every piece of evidence related to NASA's alleged Moon landings are questionable including the moon rocks, photographs, film footage, spacecraft, and related equipment specifications to name a few.

Which explains why NASA continues to hide most of the Apollo Moon landing evidence from the public. It would only open more questions about their alleged Moon landing claims that are still based on speculative evidence only and have yet to be confirmed scientifically. One has to wonder why the American Government officials don't just admit the Moon landings were faked. It really would not be that big of a deal, and people would understand why it was done. We have seen that the American Government General Accounting Office already admitted publicly most of the Star Wars Missile Defense Systems were faked, making it seem more advanced than it actually was. The attempt was to fool the rest of the world into believing we were more technologically superior than they were.

CHAPTER 4

Basic Scientific Principles Prove Moon Landings Were Faked

Chapter 4 - Part 1

Unexplainable Moon Anomalies

-Or-

Simply NASA's Scientific Deception

The Blowing Flags

The Blowing Flags

There are hundreds of concrete pieces of evidence proving the moon landings were faked, and here is one of the easiest of all to understand. Take a look at the pictures below and ask yourself what they all have in common? The American flags are all blowing in the wind on the Moon's surface. This is one of the more ridiculous claims NASA wants everyone to accept without questioning the authenticity. In many Moon mission pictures and videos the American flag appears to be constantly fluttering and blowing around. Flags fluttering and rippling in the wind would not be possible on the Moon surface because we know there is no air on the Moon and therefore no wind.

Studio Fans Were Causing Flags to Blow

Flags blowing are exactly what you would expect to see on a movie film stage that would need huge powerful fans constantly blowing on the actors to keep them cool in their space suits. This is illustrated in the NASA picture to the right. A large black fan is seen being used to cool the astronauts. These types of oversized fans would have kept the flag blowing constantly as well.

Large Movie Stage Fan
ap11-s69-31152

Evidence suggests flag blowing was a big problem for the NASA Moon-landing hoax from the beginning.

From the first Apollo mission on the Moon we see the American Flag blowing in the wind or laying flat and dangling from the top of the pole as seen in the pictures below.

Later we see that NASA used a special horizontal metal stiffening rod on the top of the flag and through the center to help compensate for the fan wind. They continued to use this method even though it was never very effective. The flag was almost always seen rippling uncontrollably in the wind.

a17/20117331 Cross member

Basic scientific principals prove a flag simply cannot wave in a vacuum where there is no atmosphere to provide wind. Not even slow movement in a gravity-free atmosphere could cause anything like this. This is a significant clue that the moon landings were staged in Earth's atmosphere. Even NASA officials confirm the flag would not have been seen blowing in any wind on the Moon.

The Apollo 17 alleged Moon landing film footage shows the best footage of the flag blowing problems. In the scene below, one of the astronauts stated, "The flag appeared to be blowing around in the wind." This reference to the astronaut claiming the flag is being blowing around on the Moon surface can be found in the NASA clip A17v.1182126.

From NASA film clip A17v.1182126

MOON ROCKS ARE FROM EARTH

Evidence the Moon Rocks Are From Earth

Another question that needed to be answered was how NASA got the Moon Rocks if they did not go to the moon. There are samples totaling 382 kilograms (842 pounds), comprising of 2,196 individual specimens. These specimens have been processed into more than 97,000 individual catalogs. NASA out-did themselves collecting over 800 pounds of meteoroids. They could have just collected a few pounds and provided a sample to the public and made up the amount of the rest of the rocks. No one could have ever proved how much rock NASA obtained.

After careful examination we determined that NASA's Moon rocks records are unquestionably unique, and differ from traditional Earth rocks in many ways. The Moon rocks presented by NASA are between 3.1 and 4.4 billion years old and evidently contain minerals such as unaltered olivine not common in Earth rocks.

If these specimens are indeed from the Moon, then how did NASA get them if they never sent men there? NASA could not have obtained genuine lunar surface material the same way the Soviet Union used an unmanned space probe. The Soviets were far more advanced than the U.S. at the time and only collected about ten ounces of material from the moon surface after many tries. This effort occupied a major portion of the Soviet space program's capacity. It would have required a similarly large effort by the U.S., although records do not suggest any such attempt. There is also a big difference between the Soviet's ten ounces and NASA's claims of returning more than 800 pounds of material from the Moon.

NASA could not have artificially manufactured these rocks from scratch in a laboratory because the technology to do so was not yet developed in the year 1969. If NASA had attempted to manufacture these rocks they would not have been able to fool scientists around the world. The NASA Lunar Sample Laboratory has distributed these rocks all over the world. Museums such as the Smithsonian let the public touch and examine rocks they claim came from the Moon. There are many scientific books written about the Apollo Mission Moon rocks. Scientists and geologists in dozens of countries in hundreds of research facilities have examined the Apollo Moon rock samples. To this day none of them have challenged if they were actually moon rocks!

Why none of these scientists have ever questioned NASA's claim of bringing these rocks back from the moon only time will tell, it such a no brainier. To even suggest anyone could touch a rock from the moon is the biggest clue of all that NASA moon rocks could not have been collected on the moon. Unlike meteorites found on earth, which have burned off most of their radioactivity as it is heated up while passing through earth ozone layer, a rocks brought back from the moon would be highly radioactive. These moon rocks radiation levels would be deadly to extremely harmful for humans who came in contact with it for thousands of years. If you have ever seen a superman movie; a Moon rock would be deadly to humans as green kryptonite was to superman. So if you humans have yet to collected rock directly for outer space and returned to earth, where did NASA get their moon rocks?

NASA simply had geologist's search for a large group of meteoroids rocks that had fallen to Earth and claim them to be Moon rocks. Unsuspecting scientists around the world determined the rocks were without a doubt from outer space and assumed they were from the Moon, since they had all seen the Moon rocks collected by the Apollo astronauts on live television.

Determining how NASA got the Moon rocks is really no deal after understanding that their goal in faking the moon landings was only to fool everyone for a short period, just long enough to keep the World safe through the cold war era. The reason no one suspects NASA's method of fooling everyone is because it is so simple. They always assumed eventually their hoax would be discovered and in their wildest imaginations never expected it to last this long. No one working on the project ever thought NASA would be forced to conceal their Moon hoax forever. The American Government's hoax was just a ploy to fool the world and make people believe they had technological superiority over the Soviets. It was important for the U.S. Government to steer the rest of the world away from leaning toward becoming a communist style government.

NASA Moon Rocks Were Counterfeits

Beside the fact that NASA's Moon Rocks are missing the high levels of deadly radiation there many other clues they are counterfeits! Yes, all these years NASA has been pulling a fast one on everyone. They have been passing around doctored up meteorites as real meteoroids.

Not many people realized the difference between a rock that is traveling through outer space called a "meteoroid", and a rock from outer space that has landed on earth called a "Meteorite". Once a space rock, traveling through space enters Earth's atmosphere, it loses some of properties in addition to burning off most of it radiation and becomes what is called a meteorite. There is no real mystery as to where NASA got the moon rocks. They were simply meteorites NASA collected on Earth. If you did not get that last moon rock analogy, don't feel bad no one else did either.

Moon Rocks From Antarctica

Wernher Von Braun in Antarctica (NASA)

Here again, the famous movie producer Wernher Von Braun was also in charge of NASA's moon rock project. Reports have been uncovered that, in 1968, Von Braun and his team of geologists spent almost six months looking for meteorites in Antarctica. NASA found out that Japanese scientists discovered a large number of meteorites in Antarctica. Further, the scientists believed the rocks came from the moon. NASA simply picked up all 800 pounds of the rocks, and claimed they were from the moon. It was that easy!

The meteorites, discovered in Antarctica, were identical matches in origin. They had the same original polished micrometeoroid bombardment and exposure to cosmic rays that the moon rocks had. To compensate for the smoothing of the rocks, all NASA had to do was break them up. Once they were broken up, NASA zapped small holes in them with a Beebe pellet sprayer (little metal marbles). For this reason, the rocks appeared to have been bombarded by micrometeorites. Next, NASA simply enriched rocks with a little helium-3 radiation, which can be found on Earth, or manufactured by mixing Deuterium with Deuterium. NASA's budget made this an easy task. After that, no one could tell the rocks were not from the moon, because no rocks have ever been collected from outer space.

Ask yourself what is more realistic. NASA sending 12 men to the moon to collect 800 lbs of rock, or shooting Beebe pellets into some rocks here on earth?

Here is the funny part about NASA's alleged moon rocks. NASA claimed their rocks were from the Moon because they matched the Moon meteorites found on Antarctica. When NASA was asked how they knew the meteorites on Earth were from the moon, they said because they matched the rocks the astronauts brought back from the moon. NASA evidence of proof that the moon rocks are real is based on an old riddle that has no answer. You may have heard some if these types of riddle before such as;. Which came first; the chicken or the egg?

Their explanation of the true origin of the moon rocks is just another example of how NASA tried to keep everyone feeling unsure about the subject. Remember, keeping everyone confused was the only way NASA could continue to keep the truth about their fake Moon landings a secret. Since no one questioned the ridiculous "ring-around" explanation, it really makes a person wonder if maybe the American Government is slipping a little something into the public drinking water. By doing this, the government can control people's thinking pattern. Nothing should surprise us. Ever since President Kennedy was murdered, the American Government has mysteriously gone wild.

What about the Soviets, they had ten ounces of real moon dust that would prove the U.S. was lying about their supposed Moon rocks. Why didn't the Soviets challenge NASA's claims of traveling to the Moon and collecting samples? Evidence suggests the Soviets were terrified that if they confronted NASA it would cause a backlash of their own space program. So they went along with NASA's Moon landing claims.

Chapter 4 - Part 2

Deadly Radiation In Deep Space Would Have Killed The Apollo Astronauts

More about the Deadly Radiation of deep outer space.

Most of the kids and adults were intrigued about how outer space is filled with deadly radiation that emanates (comes) from solar flares firing out from the Sun. They were shocked to learn that there was less than a 1 in 6,000 chance of getting a manned ship through the radiation belts to the moon and returning them safely to Earth. The odds of success were unacceptable by any standards. If the Casinos gave odds on the astronauts making it to the moon and back, a $100 bet would pay you $600,000 if only one of the missions were successful. Just imagine if you could be as lucky as NASA claims to have been! Now multiply that by six missions making it to the moon and back safely. Your $100 bet would make you one of the richest people in the world!

Many were unaware of Earth's radiation belts, and how they protected the astronauts orbiting Earth. An example would be like those who recently fixed the Hubble Telescope and those who worked on the space station. They didn't realize that none of the manned space missions, other than the American's Apollo Moon Landings, ever traveled more than 560 kilometers (358 miles) from Earth. They were shocked to learn that NASA, some thirty plus years later, now claimed that the technology to penetrate Earth's radiation belts no longer exists. If that's the case, their claims of sending the Apollo Astronauts to the moon, back in the 1960's and 1970's, is simply nonsense. Could NASA really have this secret technology no one else can seem to get their hands on? Obviously not, if they did, then why has every other NASA mission, in the last 30 years, been forced to stay well within the safety of Earth's orbit?

Even President George W. Bush, on January 17, 2004, stated NASA does not have the technology to travel more than 358 miles above Earth. Furthermore, some new information indicates that NASA just completed a top-secret experiment. Hence, my theory about the radiation belts is confirmed. An unnamed source claimed the Columbia Space Shuttle attempted to approach the first radiation belt. At approximately 560 kilometers (350 miles) from Earth, the astronauts turned back. Moreover, they claimed the space shuttle cabin lit up completely and the radiation was so strong that they could see lots of little lights through their tightly closed eyelids.

When we confronted NASA, they would neither confirm nor deny this testing. One can only imagine there is now a new NASA task force in place trying to come up with a solution on how to tell the public why they felt it necessary to fake the six Apollo moon landings. Especially as more and more people question the Radiation Belt experiment as a contributing factor to the Columbia Shuttle's malfunction that caused it to blow up over Texas killing all seven astronauts.

In addition to being exposed to deadly radiation, the Apollo astronauts would have been pierced by thousands of micrometeoroids which are scattered around solar systems. One would have expected the micrometeoroid bombardment and extremely deadly gamma and cosmic radiation would have killed the astronauts within the first few minutes after attempting to leave earth's inner orbit.

NASA's own administrator Sean O'Keefe confirms Hoax theory

Possibly the biggest slip of the tongue that confirms the 560 kilometer (350 mile) distance limit was made by NASA's own administrator Sean O'Keefe. He recently told 3,800 people gathered for the 18th Annual National Space Symposium that NASA just spent $180 million on research to penetrate the Radiation Belts. He said NASA is dedicated to solving the problem of extreme radiation exposure that could halt any human space exploration. Did NASA's own administrator forget they already claimed to have penetrated the radiation belts many times on the way to the Moon back in 1960's and 70's? Maybe he thought Earth's radiation belts came after traveling past the moon and not before.

The lethal radiation belts are just one example of evidence that technology clearly did not exist in 1969, and undoubtedly, does not exist today. The Apollo spacecraft was plagued with onboard guidance defects and more. In the 1960's, the technology simply did not exist to build a functional guidance system for the Apollo Lunar Module. The Apollo guidance system would have to have been built with commercially integrated circuits that would have had the equivalent computing power of a small calculator or wristwatch. Because of this lack of power, the astronauts would not have made it to the Moon.

At the time this book was written NASA had yet to build a spacecraft that could land and take off, the same as the Apollo Spacecraft supposedly did. More details on how the LM spacecraft truly functioned can be found in later chapters of this book.

If we were to recalculate the odds of NASA successfully sending a man on the moon and returning him safely based on what we know today about the levels of Cosmic radiation, the odds would be well over 1 in 600,000. Since the Nuclear bomb testing has now created a much stronger Inner radiation belt. There was virtually no chance the untested Lunar Module could have landed and returned flawlessly six times. After carefully studying NASA's design specifications on the prototypes supposedly tested here on Earth, all indications are the Lunar Module could never have managed to lift off the ground correctly, let alone fly and navigate under it's own power.

Deadly Radiation In Deep Space Would Have Killed The Apollo Astronauts

Many people don't realize that a Nuclear Powered Magnetic Force Field surrounds the earth and protects everyone from deadly space radiation. This field is called the Magnetosphere. Earth's Magnetosphere is a lot like a force field of a Star Trek Spaceship seen in the movies. This field has its shields up at all times. It is located 650 kilometers (400 miles) above Earth and extends an additional 65,000 Kilometers (40,000 miles) out towards deep space. To get to the moon, astronauts must travel through Earth's nuclear radioactive force field because the moon is approximately 240,000 miles beyond the force field.

Back in the 1950's and 1960's, there was plenty of evidence suggesting that NASA knew it was impossible for humans to travel through the Magnetosphere and beyond Earth's immediate orbit. One of the engineers in charge of technical publications and advanced research, at Rocketdyne Systems from 1956 to 1963, estimated that there was only a 1 in 6,000 chance of surviving a trip to the Moon. This took into consideration the radiation, solar flares, and micrometeorite impacts the astronauts would have experienced. This did not take into account one of the worst solar flares ever recorded between the Apollo 16 and 17 missions in August of 1972.

Furthermore, back then, the Soviet Union supposedly attempted space travel beyond Earth's orbit and the three cosmonauts were killed as they started to travel through the deadly radiation field that surrounds Earth.

News of the Soviet disaster spread quickly, and NASA began to wonder how their astronauts were going to travel through the radiation field. Then, they found a solution to the problem, and it was a biggy. NASA engineers theorized that a nuclear bomb explosion would vaporize the radiation particles. Just as they would clear a passageway through a minefield. As a result, the Apollo Astronauts could travel freely to and from the moon. Therefore, the American Government detonated a massive Megaton Nuclear Bomb approximately 257 miles above Earth, in 1962, in an attempt to try and force an unnatural passageway through the Van Allen belt. The huge nuclear detonation was disastrous! It had the opposite effect and the radiation level increased so dramatically that there was no way the astronauts could get through.

It was Earth's Magnetosphere that prevented the astronauts from traveling any further out to space. The Magnetosphere is filled with extremely deadly radiation from outer space. This shield blocks large amounts of extremely deadly radiation from reaching Earth's surface.

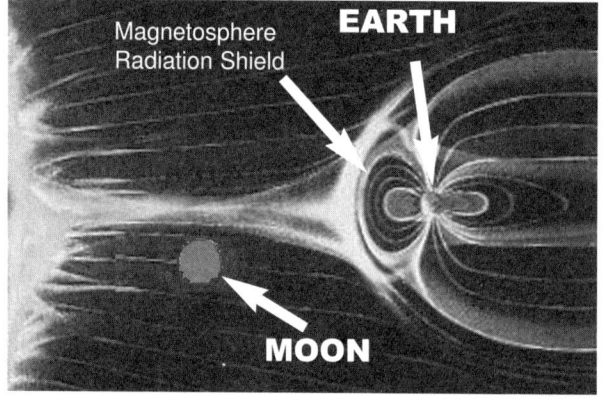

This force field has the ability to absorb huge amounts of radiation and reflect radiation back into outer space. Without this force field surrounding earth, the deadly radiation from outer space would quickly kill everything living on the planet. When the Sun has a solar flare, it's like detonating 100 nuclear bombs all at one time. This explosion, on the sun, causes very large amounts of deadly radiation, at temperatures over 500° million degrees Fahrenheit or 300° million degrees Celsius, to start racing towards Earth.

Once the deadly radiation begins reaching the outer perimeter of earth's protective force field, it bombards Earth with large amounts of Alpha, Beta, and Gamma X-ray radiation. The radiation strength can exceed 200 MeV and temperatures over **269,540° F or 149,726° C** are common.

To illustrate just how powerful these solar flares from the Sun can be when they reach Earth's protective force field, an x-ray, taken in a doctor's office, can be used for comparison.

At a radiation level of 1 MeV in gamma radiation, a high power x-ray machine takes a picture of bones. Therefore, solar flares from the sun on the way to the moon and on the surface are 200 times more radioactive than an X-ray taken at a doctor's office. Hence, the flares can last for hours at a time. During an individuals entire lifetime, the most he or she is supposed to be exposed to is 1 MeV of radiation for 2.5 seconds.

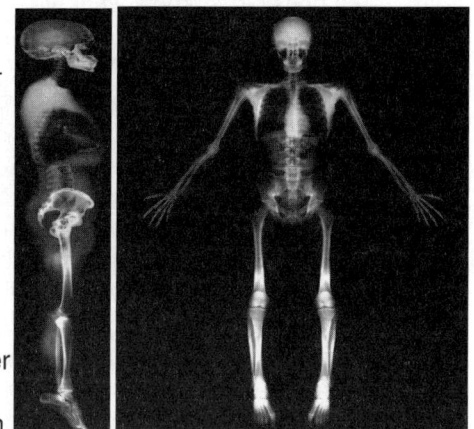

There is absolutely no way the astronauts could have survived on the moon surface receiving such high dosages of radiation for such a long period of time. In addition to the deadly gamma radiation, the Apollo Astronauts would have been subjected to an even more deadly form of radiation called cosmic radiation. This cosmic radiation comes from other galaxies in deep space and is produced by things like exploding stars. Cosmic Radiation has been measured over a million times stronger than an X-ray at a doctor's office at 2.7 TeV. This means the Astronauts would have been exposed to x-rays over one million times stronger than someone would receive for something like a broken leg. Again, with the windows installed into the spacecraft and the helmet visors made of glass, the Apollo Astronauts would have had no protective shielding from the radiation and died instantly. For more details on the radiation exposure the astronauts would have been subjected to, read the advanced theories chapter of this book.

Imagine if Earth's surface was openly exposed to this deep space radiation as the moon is. Where would you hide? Every living thing on the planet would be vaporized. The question that NASA needs to answer is the following: how did the Apollo Astronauts manage to survive the solar winds and cosmic radiation they would have been exposed to in deep space? NASA can't answer that question, and they don't because it would expose their fake moon landings.

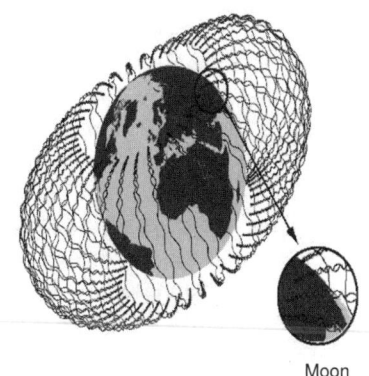

Moon

The secret to NASA's Moon landing hoax going unchallenged for so many years is because they are spending billions upon billions of dollars trying to keep everyone confused about the issue. For instance, NASA has been claiming all these years the astronauts were only exposed to a small amount of radiation. However, they only reference the alpha and beta radiation. They also avoid discussing the third type of radiation in outer space gamma x-ray, which is the most deadly radiation of all. If NASA were to even comment on the cosmic gamma radiation, it would instantly expose their alleged Moon landings as being fakes.

In space, gamma radiation can be millions of times stronger than the alpha and beta radiation and would have instantly killed the astronauts the moment they came into contact with this type of radiation. Gamma radiation is the most penetrating of the three basic types of radiation. With Gamma radiation, humans need a shield to protect themselves and most substances like paper, skin, and wood, will not stop the radiation from penetrating the body. This type of radiation causes severe damage to the human body. It starts with nausea and vomiting, then internal organ damage, cataracts, and ends with cancer eventually killing the person. The good news is, in it's natural form, gamma radiation is mostly found in space.

What Would Be the Required Amount Of Shield Protection For the Apollo Astronauts To Travel To the Moon.

It is possible to know what type of protective shielding would be required for the astronaut's journey to the moon. Yes, it is possible to determine exactly how much protection the astronauts would need. However, it involves a very lengthy formula. The actual formula can be referenced in the Advanced Theories Chapter of this book. Instead of having to go through each step of this formula, here is a shortened version.

For this illustration, it is much lower than what the actual radiation level the astronauts would have been exposed to, we will use NASA's estimate of 1 MeV of radiation. According to the nuclear regulatory industry standards, 1 MeV of gamma radiation would require the spacecraft to be protected by a shield of either 5-inches of steel or 1 inch of lead. Nonetheless, the Apollo spacecraft was only equipped with a very thin layer of aluminum foil, a 1 inch steel wall, and the astronaut's space suits. Wrapping the spacecraft in aluminum foil would have only protected the astronauts against approximately 30 MeV of the very weak alpha and beta type of radiation, but would be ineffective against 1 MeV of gamma radiation.

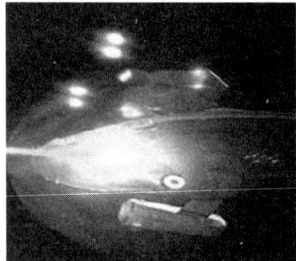

Another problem with NASA's claim was the glass windows in the Apollo LM Spacecraft and command module, provided basically no protection against any of the different types of deep space radiation. When NASA's engineers installed windows into the spacecraft, it's hard to imagine what they were thinking because they left the astronauts with basically no protection. This Apollo spacecraft surely looks as if it were designed for movie purposes only.

The 1-inch lead shielding was based on NASA's original estimates made back in the 1960's of only 1 MeV of gamma radiation. Although, according to more recent measurements of the cosmic radiation outside earth's protective force field, the Magnetosphere power levels are actually in the trillion strength levels. These levels are referred to as TeV. At the TeV strength, the astronauts would have needed a ship with a lead shield much thicker than originally thought. If these modern day instruments are correct, then NASA is clearly lying about sending men to the moon back in the 1960's and 1970's. This would also explain why the space shuttles cannot travel more than 400 miles into space and must stay within earth's protected inner orbit.

Houston we have a problem! If you are planning to use a primitive mineral based protective shield such as lightweight lead for your new spacecraft, it's not going to work. In order to transport humans to and from the Moon, without being vaporized by the cosmic radiation of deep outer space, the minimum size of your twenty first century spacecraft would be a spherical ship with an exterior lead shield in all directions many miles thick. Seems a little impossible. Don't you think? How are you going to ever get it off the ground?

NASA must know this and their 2020 moon landing is just another attempt to filter money out of the American taxpayers pockets. Perhaps they have managed to actually develop a Nuclear Diffusion Shielding System to block space radiation, which includes a Particle Disrupter for those marble sized meteoroids that can really cause some serious damage to a spacecraft.

The natural forms of deadly space radiation were not the only type of radiation the astronauts would have needed protection against. Furthermore, what about the nuclear radiation from the bombs NASA used to clear a pathway through the magnetosphere? Between 1958 and 1972, the American Government had exploded many hydrogen and nuclear bombs into Earth's inner orbit. This increased the level of radiation the astronauts would have been exposed to.

Since the 1950's, this type of nuclear weapon testing has continued and current readings of the inner radiation belt indicate that the levels of radiation have increased by the millions and are out of control. To date, there has been over 2,000 nuclear detonations in outer space that are not only dangerous to astronauts, but this man made radiation is also destroying Earth's protective force field.

In 1958, before the United States started exploding nuclear bombs in outer space, the Explorer 1 was launched into orbit carrying a geiger counter to record radiation levels. It registered nothing at 2400 kilometers (1500 miles) altitude. Several months later, after NASA exploded nuclear bombs into Earth's inner orbit under the code name Argus (summer 1958), Explorer 5 was launched. It also carried a geiger counter, but this time it was encased in one centimeter of lead. This time there was a huge cloud of radiation hovering over the planet at 600 miles above earth, which was clearly not there before.

If you think it sounds like the United States created this deadly inner radiation belt, your not the only one wondering about that. The Soviet Union, since the 1960's, has been accusing the United States of making the deadly inner radiation belt by exploding nuclear bombs above Earth. It's more likely that the belt was always there, but was not storing any deadly radiation. Once the nuclear bombs exploded into outer space, the belt started to fill up.

All the American's managed to do with there nuclear bomb testing was force the Soviet Union to reduce their working space from a 1500 mile altitude to a 400 mile altitude. In order to stay away from the new deadly radiation cloud, this is now necessary.

Until now, government officials have been cleverly covering up the fact that they are the ones who have been charging the inner radiation belt with nuclear radiation and for good reasons. The effects of these nuclear testings are catastrophic to earth. It is already destroying Earth's ozone layer resiliency.

The Soviet Union knew a trip to the moon was impossible. It's been reported they secretly sent manned spacecrafts looping around the far side of the moon at least twice in 1968, under the codename "Zond".

Soviets Zond Five Manned Spacecraft

On September 14, 1968 the Soviet Union launched "Zond 5", which carried one cosmonaut and some smaller earth creatures including turtles. The spacecraft was also equipped with an on board voice transmitter to relay voice sounds back to earth. The purpose of this "Zond Five" flight was to circle the moon and return to earth safely. The Jodrell Bank observatory in England reported they had been carefully monitoring the voice transmissions of the cosmonaut during the mission. They claim the cosmonaut's transmission only lasting a short period, before going dead. When Zond 5 cabin was successfully retrieved from the Indian Ocean at the end of September 1968, the cosmonaut and all the earth creatures were rumored to have been completely vaporized.

This picture was taken of the actual turtles as they were being placed on the Zond 5 ship to the moon

Under pressure resulting from the United States claims of launching the three Apollo 7 astronauts on October 11, 1968 to the moon and returning the men safely, the Soviet Union made one more attempt to be the first on the moon and sent three cosmonauts on a suicide mission to the moon, on November 10, 1968 under the codename "Zond 6". The Jodrell Bank observatory again monitored the cosmonauts voice transmits with similar results to the previous Zond Mission. The Zond 6 cosmonauts' communications only lasted a short period, before going dead.

Zond 6, during capsule re-entry, is reported to have crashed to the ground because the parachute did not deploy. The Soviets were disappointed and immediately canceled all future programs to send men to the moon from that day forward. Some suggest that the Soviets also immediately notified the American government, and NASA, of the Soviet's Zond 5 and Zond 6 test results. When the American government explained to the Soviets that they were only faking the Moon landings to please the American people during troubled time, the Untied States asked the Soviet Union's top ranking officials for their cooperation in keeping their moon hoax a secret. After agreeing on the appropriate amount of financial concessions the Soviet Union would receive of the American taxpayer's money the treaty was signed.

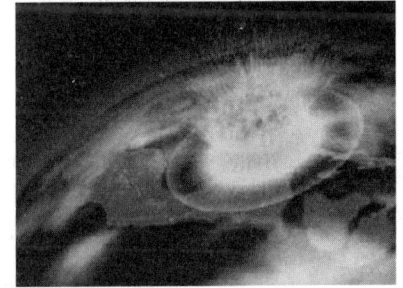

Nuclear Bomb Explosions Cause Holes In Earth's Ozone Layer

Contrary to what the American Government wants everyone to believe, pollution is not the primary cause for the ozone layer holes in the North and South Pole regions. A majority of the ozone layer depletion is a direct result of the secret nuclear bomb testing in outer space, which has caused a dramatic increase in nuclear energy in Earth's Inner Radiation belt. When a nuclear bomb is tested in earth's orbit, it does not drift out into space. It becomes trapped by the inner radiation belt and gradually drifts back down to earth. While the bomb is moving back to earth, it eats away at the Earth's ozone layer. With over 2000 nuclear bombs already tested in outer space, the ozone layer simply no longer has the ability to hold up any longer and it is decaying. More detailed explanations of these nuclear testings are discussed later in this book and it's worth taking note.

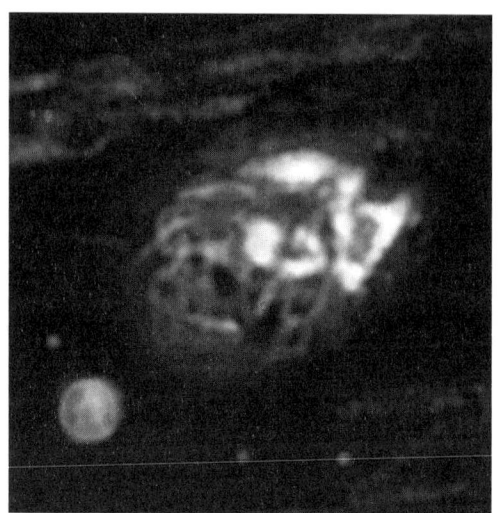

The side effect of the increased unnatural nuclear fallout from the bomb testing has already started the erosion of the Earth's ozone layer. Just as would be expected, the side effects are beginning to show up in the aquatic reptiles species first. This is because, in many cases, these creatures have much thinner skin than humans. Hence, these reptiles require a much smaller time frame than humans for side effects of prolonged exposure to begin to manifest. Unfortunately, the end result for all species, living on Earth, will be the same.

If you think this is impossible, then check it out for yourself. Independent research around the world is already beginning to show signs of deformations in aquatic creatures at an alarming rate. The creatures with the thin unprotected skin like frogs are the first to be affected. For God's sake, look what is already happening to Earth's beautiful sea turtle. Their bodies, in many cases, are almost completely covered with large cancerous tumors. As a result, the turtles eventually die. Many Turtles, with these Fibropapillomatosis tumors, can be found in Florida, Japan, Barbados, Australia, the Pacific coasts of Mexico, Costa Rica, and everywhere else. In some parts of Hawaii, it is estimated that the tumors are present in 60% of the turtles. In addition, these Fibropapillomatosis cancer tumors interfere with the reptile's ability to feed, eat and swim. Because a few governments want to protect themselves with nuclear weapons, can you imagine 60% of the human population being affected in this manner?

In the upcoming Ice Age, there is one big difference. The Human Race is threatened by extinction, rather than dinosaurs. The first problem facing humans is, since the last major Ice Age they have migrated to all regions of the globe, most of which cannot sustain human life during an Ice Age. Also, the population has grown from less than 50,000 humans living on the earth during the last Ice Age to over six billion today. Most humans will perish during the first 300 years and the remaining 50 million or so people will be fighting for their lives due to the limited resources that will be available during the worst conditions this Ice Age will bring. This opinion is shared by many other scientists around the world who are terrified by the environmental side effects of nuclear bomb testing. Americans are especially concerned since they were the ones to start it.

One attempt NASA is taking to stop the potential ecological disaster of a deteriorating atmosphere is constructing the HARP, which stands for High Atmosphere Research Project. Many believe this HARP Project is trying to make an artificial ozone layer barrier to patch the large holes created in the North and South Pole regions. Others believe NASA is about to once again show how brainless they can be, and try to open a hole in the earth's protective force field the magnetosphere. This time it is not to send men into outer space. There is not one educated geophysist that honestly believes sending men to the moon is a possibility. The reason they would be trying to open a hole in the Earth's protective force field, the magnetosphere, is to try and let all the nuclear radiation out before it collapses back onto Earth and kills everything on the planet. If this rumor is true, then the Harp project needs to stop immediately. Once the magnetosphere is opened up, the radiation will not leak out into space; it will more than likely create a black hole effect drawing nuclear energy into Earth's inner orbit vaporizing everything on earth's surface. Next, the radiation will travel along the magnetic flux lines to the center of Earth's core and explode the planet into trillions of pieces. Oh, this secret Harp project should not be confused with the Harp cannon missile test developed in the 1960's.

I apologize for getting a little off track. Lets take a look at those "high-tech" astronaut space suits of the 1960's.

THE SPACE SUIT

Apollo Astronauts Space Suits Lack Adequate Protection

Could those astronaut space suits have really protected the astronauts from the deadly radiation levels of outer space? As mentioned before, the helmets with glass visors didn't provide enough protection against any of the deadly radiation that emanates from the solar flares firing out from the Sun and often bombards the moon's surface. During the alleged Apollo Missions, astronomical data shows there were no fewer than 1,485 radiation solar flares. Additionally, temperatures reached over 269,540° F or 149,726° C. Yet, none of the Apollo Astronauts were burnt to a crisp!

If NASA had such a powerful space suit material, it would be considered one of the greatest technological breakthroughs of humankind. These super suits could protect rescue workers standing on top of a nuclear power plant as it was melting down. They could go about their job without worrying about getting cancer, since they would only be exposed to a fraction of the lethal radiation the Apollo astronauts were exposed to during their mission to the Moon.

Other interesting facts about the space suits include:

The astronaut's space suits were not designed to provide adequate protection from the extreme lunar (moon surface) temperatures, which reaches 250° F in sunlight, and below -250° F in the dark. There was not even sufficient protection for the huge temperature contrasts between the sunlight and the shaded areas of the Moon.

It would have been impossible for the water cooled space suits to work on the Moon when outside temperatures were already at the boiling point of water. The lunar surface gets continuous daylight for two weeks at a time and there's no atmosphere to carry the heat away. In a space vacuum without enough surrounding material to dissipate the heat the astronauts' would, have been baked alive in the sun light and frozen solid in the dark.

If NASA's designers in the Crew System's Division, as well as Hamilton Standard, who manufactured the suits, would release classified technology specifications related to the space suit. Then a team of independent scientists and engineers could confirm or disprove NASA's claims about the suit's capability using modern day technology.

What about the astronauts' glass visors installed in the helmets.

They would have provided no protection against the solar winds and galactic radiation bombarding the astronauts while standing on the moon surface.

All the Pictures On This Page Have One Thing In Common; the Astronaut's Space Suits Are Torn.

Tears In Astronaut's Space Suits

Imagine if these were actual Apollo astronauts seen in these Moon landing photos. With holes like these in their space suits, they certainly would not have survived the airless conditions of outer space.

NASA ID# as17-134-20380- Slit In Pants

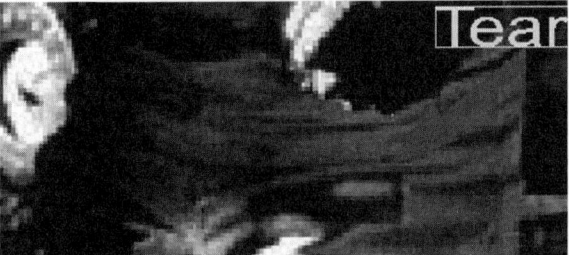

Close-Up of torn area in space suit

Large Rip in the left leg of Apollo 12 Astronaut's Space Suit while he was supposed to have been on the Moon

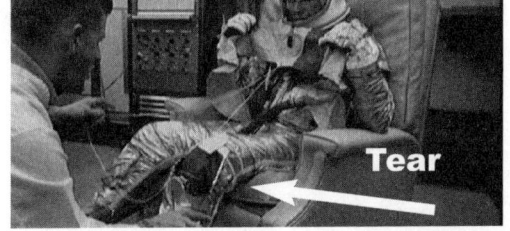

4-21

Tears in astronauts' space suits -continues

Oh look here while working in space, one of the Apollo Astronauts accidentally step on the other astronaut and rips his space suit wide open. Spaces must have been a tricky place back in the old days, when astronauts had to maneuver around all those levitation harnesses supporting the ship.

NASA ID# 10076104

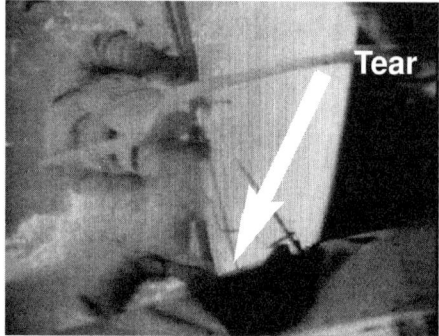
NASA ID# 10076107

NASA Officials Confirm Trips to the moon IMPOSSIBLE

Several NASA officials have made comments that seem to support the Hoax Conspiracy theory. Several have made comments indicating that humans cannot travel more than 560 kilometers (350 miles) above Earth. Listen to what a few other NASA officials had to say about how long they believe humans must wait before a manned mission to the moon is possible.

President George Bush: On January 15, 2004 President Bush unveiled his goal to send men to the moon by 2020. He stated this would be a "new course" for the nation's space program in a speech at NASA's headquarters. Did President Bush not realize his new goal was over fifty years too late, and the American Government already claimed to have sent men to the moon back in the 1960's and 70's?

Of course Mr. Bush knows about NASA's original claim of sending men to the moon. To brace everyone for the bomb shield NASA is about to drop on them, this was obviously just another part of NASA's " dezincification" process underway. That bombshell is telling everyone that they faked sending the Apollo Astronauts to the Moon.

As crazy as it may sound, NASA is going to admit that they faked landing men on the Moon, and you heard it here first. Why else would the President on National TV announce that NASA has been able to send men beyond Earth's low orbit and through Earth's radiation belts? Why would President Bush tell the world NASA no longer has the technical ability for humans to travel more than 400 miles above earth?

DOUGLAS COOK: In 1999 Douglas Cook, Director of the Exploration Office at Houston's Johnson Space Center, calculated that man should be able to reach the moon within 100 years. Why would he say this if he knew man has supposedly already achieved this goal?

DAN GOLDEN: NASA Chief, openly admitted during a TV interview with UK Journalist Sheena McDonald, that humankind cannot venture beyond Earth's orbit more than 400 kilometers (250 miles) into space, until they can find a way to overcome the dangers of cosmic radiation. He must have forgotten that NASA already claimed to have sent 27 astronauts 400,000 kilometers (250,000 miles) outside Earth on previous Apollo Moon landing missions thirty-four years earlier.

The obvious reason that these NASA officials are making these statements about human's first trip to the Moon as being a long way in the future is because the technology never existed in 1969, and it certainly does not exist today. Clearly, the Apollo manned missions to the moon were fabricated here on Earth. One can only wonder if these NASA officials were instructed to release this information in an attempt to begin desensitizing the world population to the possibility that the Moon landings were a big hoax. It would certainly be to NASA's advantage to start this process well before the year 2026, when the reports are declassified. It's just a matter of time before someone develops an orbital satellite like the one that NASA claimed to use in the 1960's to orbit the moon and discover there is no evidence of Americans ever landing on the Moon. It appears this is a big concern of NASA since they are currently trying to pass an international law making the alleged Apollo landing site off-limits to unmanned space vehicles. They are essentially establishing a No Fly Zone to insure the hoax is kept a secret. Even if the International Law passes, what countries would abide by a No Fly Zone on the Moon. The truth is going to be revealed sooner or later.

Our research group consists of honest, caring, good-hearted people that feel the public has been kept in the dark long enough about the landing hoax! They realize the current NASA officials running the space program have the power to set right the errors of judgment made during those Apollo days, and make the world a better place.

This book honors all the heroic individuals who have put their lives on the line whether knowingly or not. The members of the WhizKid Project would like to dedicate this book to Astronaut Sally Ride, NASA Director Douglas Cook, and NASA chief Dan Golden. The WhizKids would also like to dedicate this book to those individuals suspected of being murdered by NASA officials and the CIA just before they were going to go public and expose the moon hoax. This list includes the many Apollo original astronauts who tested the Apollo Moon landing equipment, a reporter covering the Apollo missions and President John F Kennedy (JFK).

The lethal radiation belts are just one example of evidence that technology clearly did not exist in 1969 and undoubtedly does not exist today. Again NASA knew this as early as 1963 when NASA engineers calculated the chance of a successful Moon landing to be 1 in 6,000.

All evidence indicates the Apollo spacecraft was plagued with onboard guidance defects and more. In the 1960's, the technology simply did not exist to build a functional guidance system for the Apollo Lunar Module. The Apollo guidance system would have to been built with commercially integrated circuits that would have had the equivalent computing power of a small calculator or wristwatch. This was nothing near what would be required to go to the Moon.

If we were to recalculate the odds of NASA successfully sending a man to the moon and returning him safely based on what we know today, the odds would be well over 1 in 600,000. Since the Nuclear bomb testing has now created a much stronger Inner radiation belt. There was virtually no chance the untested Lunar Module could have landed and returned flawlessly six times. After carefully studying NASA's design specifications on the prototypes supposedly tested here on Earth, all indications are the Lunar Module could never have managed to lift off the ground correctly, let alone fly and navigate under its own power.

Confirming The Moon Landing Hoax with Radiation Experiment

Before getting into the rest of the amazing evidence, the Internet Whiz Kids want to mention their first of many scientific challenges for NASA. They have developed a simple test that will prove without a doubt whether or not NASA has the mysterious secret technology they claimed they used in the 1960s and 1970s to travel to the moon.

Internet WhizKids Experiment #1: WhizKid, Ricky (age 13), came up with this very simple test. Rick some day wants to be a nuclear physicist and gives a new meaning to the word WhizKid. Ricky, on his 2nd birthday party, was filmed going through the entire boot sequence on his computer, searching the menu screen, and then selecting the spelling phonic games he loved to play. His fourth grade state educational exam shows the score was so high on the scale, the computer rejected his high score and it had to be recorded by hand. He recently ranked the champion in a spelling bee out of all the boys in the region over 20,000 students. Furthermore, he spelled several words the speaker could not even pronounce correctly. Ricky's talents are good enough to get him hired at the age of 12 as assistant editor for this book.

Here is Ricky's suggestion, have one of NASA's manned space shuttles that is in outer space, simply leave the Earth's orbit and travel roughly 24,000 miles away from the Earth's surface. Have the astronauts take a few pictures and return to Earth without being burned to a crisp. The WhizKids insisted NASA allows an independent confirmation of the mission. They would even be generous enough to agree to let NASA modify the space shuttle to meet the original Apollo spacecraft's specifications. That would only require NASA to remove the shuttle's heat shields and leave the large cargo bay door open during the trip through Earth's radiation belts. Remember that the Apollo LM Spacecraft only had tin foil for a ship shield, so NASA may want to downgrade the shuttle's protection. If NASA possesses the technology to travel to the moon, this should be a very simple experiment for them. The distance is only one-tenth the distance to the moon and could be completed easily in one day. If NASA can perform this very simple experiment, they can prove to the world they have the technology they claim to have.

The WhizKids estimate it would cost NASA less than 17 million dollars and could be conducted immediately. The Internet WhizKids have set up a fund raiser to help NASA raise the $17 million to carry out this space travel experiment. They truly want to prove whether NASA can or can not send a manned spaceship beyond Earth's orbit without being burned to a crisp. If you would like to help them reach this lofty goal, you can make a donation at www.whizkids.tv. This experiment is very simple and if it can be fully funded there is no reason NASA should refuse to do it, other than being unable to do it. Until then, it is NASA's word against common sense and basic scientific logic.

ap15-s71-42037

Chapter 4 - Part 3

Who is really telling the truth NASA or the Billions of Stars in the sky?

Lack of Stars Proves Moon Landings were faked

The lack of air on the Moon means that outer space would appear very dark. If you were standing on the Moon you should see billions of stars, especially on the dark side. Yet, when the Apollo 11 astronauts were asked, "when you looked up at the sky, could you actually see the stars?" Neil Armstrong replied, "We were never able to see stars from the lunar surface or on the daylight side of the Moon by eye, without looking through the optics (i.e., the lunar module's navigation telescope)." Then Astronaut Collins, who supposedly remained in the command module in deep space replied, "I don't remember seeing any." These statements alone make it clear they were lying about ever going to the Moon. Reference source; The First Lunar Landing As Told By The Astronauts: Armstrong, Aldrin, and Collins in a Post-flight Press Conference, NASA EP-73, 1989 pt. VI).

No stars were ever mentioned by any of the astronauts during the alleged seven trips to the Moon. This would only make sense if the Moon landings were faked here on Earth. Remember, in 1969 people did not understand the stars or space. NASA engineers themselves could not be sure if the astronauts would see stars. Today we realize the astronauts would have had the opportunity of a lifetime to see and take pictures of many stars in the universe, visible from the Moon. They would have seen an overwhelmingly spectacular sight of billions of stars, not to mention being able to see Jupiter, Saturn, the Milky Way and so much more. This would be like seven different families taking a trip to Disney World and none of them ever noticing the rides, the food stands, or other people attending the park. They simply could not have gone to the Moon without noticing outer space.

The Apollo Astronauts Must Have Forgot What Stars Look Like!

Now the purpose of this book is not to accuse the Apollo Astronauts of being liars. Moreover, it's to prove they were scientifically dishonest.

While standing on the moon surface, all of the astronauts stuck to the same story. They stated that stars could not be seen. Because of this, all we have to do is prove they would have seen the stars from the Lunar (Moon) surface. This would confirm they had been untruthful about going to the moon.

Nevertheless, if there was no way of traveling to the moon to take pictures for ourselves, how could we prove the astronauts would have seen stars? Once again, we will rely on NASA's own evidence.

To prove the stars would have been visible, the first thing that could be used is the Apollo 16 Astronauts mission records. While standing on the Moon's surface, these astronauts claimed they could not see any stars in the sky. The Apollo Astronauts declared that they could only perceive the Sun, the Earth, and some Magellic and Magellanic Clouds.

Someone might want to explain to Apollo 16 Astronauts, John W. Young and Charles M. Duke, that volcano gases on the moon surface did not produce the Magellanic Clouds. These clouds are "STAR CLUSTERS". Yes, Magellanic Clouds are star clusters scattered across the universe, such as the Milky Way in your Galaxy. Other Magellanic Clouds, like Virgo, are comprised of thousands of galaxies and hundreds of billions of stars.

According to the Apollo 16 American Astronauts, they seen plenty of stars and star clusters similar to the picture on the right.

The Apollo Astronauts were supposedly filmed while they were taking pictures of the most spectacular looking Magellanic Clouds.

Below is a clip from the Apollo 16 moon landing scene where the astronauts mentioned seeing Magellanic Clouds, and they stopped to take pictures of them.

Since the Apollo 16 Astronauts claimed they could see Magellanic Clouds and took some fantastic pictures of them, they must have been lying about not seeing any stars. No wonder NASA refuses any independent questioning of the Apollo Astronauts. Within five minutes, the whole world would know the Americans faked the moon landings.

Undoubtedly, when NASA reads this accusation of fraud, the first thing they are most likely going to do is claim the astronauts could only see fuzzy star clusters in the sky. NASA will Insist the astronauts could not see any stars. In anticipation of NASA taking this action, in order to insure the truth be told, I have put together a little Galaxy Constellation Magnitude Comparison Chart.

This chart is designed to help answer the following question. If the astronauts did see star clusters, like the Milky Way, would they have seen any other items that would be brighter in comparison? The Answer is yes! The list includes various objects, such as certain planets that are much brighter in comparison to the brightest of all Magellanic Clouds.

While still providing enough information to prove the Astronauts would have seen the stars in the sky, every effort has been made to make it as simple to understand. This chart is going to determine the level of brightness between the objects the Apollo 16 Astronauts claimed to see from the lunar surface and the ones they claimed they could not see. It really was a very simple process. In astronomy, the brightness of any star is measured using the "Visual magnitude". One method of recording the brightness of stars, which was invented by one of Ann and Joachim's friends named Hipparchus, many years ago. Even though this system was not the most sophisticated way of measuring the brightness of stars, many astronomers around the world still use the Hipparchus system.

It is a very simple system that measures the visual spectrum of brightness humans can see with their own eyes. Its simplicity makes this method of measuring stars perfect for book illustrations.

Here Is How the Visual Magnitude Scale Works.

Each object, in the sky, is assigned a number based on it level of brightness. The smaller the number an object is assigned, the brighter that object is. The larger the number an object is assigned, the fainter it will be, and the harder for you to see it.

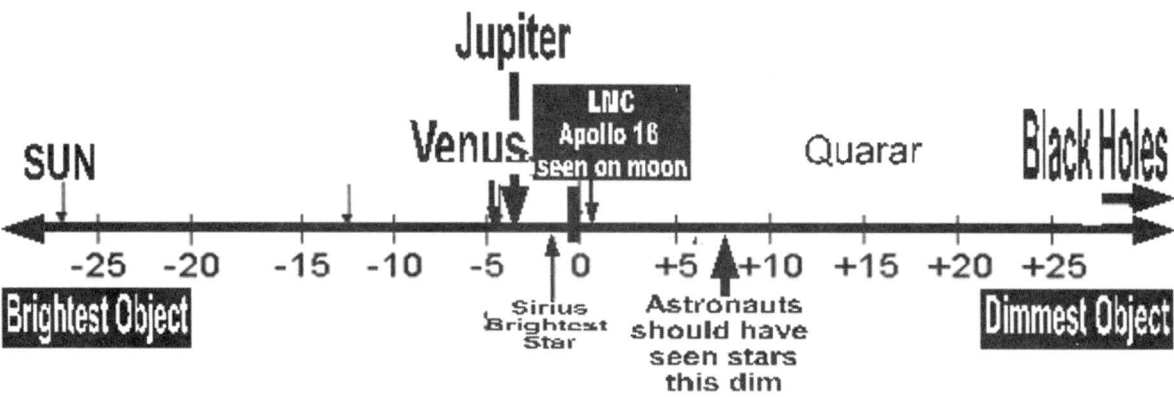

For example, the Sun, being the brightest object in the sky, would be assigned the smallest number. In contrast, a black hole would be assigned the highest number because it is the darkest. With all that said, let's figure out what the astronauts should have seen on the moon.

First of all, if the Apollo 16 astronauts were telling the truth about seeing Magellanic Clouds (Star clusters) from the Moon's surface, then we can use that as a point of reference as to what they could and could not see. In Earth, the brightest of all the Magellanic Clouds universe is Dorado/Mensa. This could has a magnitude rating of 0.1 mag. This means the astronauts must have been able to see any object in the sky with a visual magnitude rating of less than +0.1 mag. However, realistically speaking, the astronauts should have been capable of seeing many stars with a magnitude rating as high as 7.6 to 8.0 mags.

4-30

If the astronauts documented claims of seeing Magellanic Clouds were true, below is a list of a few other items they must have been able to see from the Moon's surface.

The list is in descending order from the brightest objects in the sky to the illumination level of the brightest Magellanic Cloud Dorado/Mensa.

Object In Space Brighter Than the Brightest Magellanic Cloud

Name	Object	Magnitude Rating
Sun	Sun	-27.0 mag
Venus	Planet	-4.1 mag
Jupiter	Planet	-2.8 mag
Sirius	Brightest Star	-1.4 mag
Canopus	Star	-0.6 mag
Mercury	Planet	-0.5 mag
Arcturus	Star	-0.1 mag
Alpha Centauri	Star	-0.1 mag
Dorado/Mensa	brightest Magellanic Cloud	+0.1 mag

The Picture NASA Does Not Want You To See

Below are two images taken from the moon in 1994 by the Clementine spacecraft, which show what the astronauts would have been able to see. The large white ball on the far side of the moon is not the SUN it is Earth, reflecting about 30% of incident sunshine. The smaller bright light is the planet Venus that shines above as the solar corona peaks. And of course the hundreds of small lights are stars, it looks like possibly the big dipper between earth and Venus. Again if the astronauts had gone to the moon as they claim, then this is what they should have described seeing while there.

These next pictures taken from the moon by the Clementine Spacecraft in 1994 confirm the Astronauts would have seen stars:

NO Stars in the pictures

Another important piece of evidence that helped prove the Moon landings were faked is the absence of stars in the Apollo mission pictures. After talking to several film developers and providing them with the NASA camera equipment features, they confirmed the Hasselblad 70mm Hasselblad Electric Camera the astronauts supposedly took with them should have had no problem capturing stars. Below are a few samples of the pictures supposedly taken of the astronauts working on the Moon's surface. None of them include any stars in the dark lunar sky. They should have appeared in great numbers throughout the photos taken on the Moon.

None of the pictures show any stars, not even ONE. How could this be and what was NASA thinking? Not even the closest star, the SUN shows up! NASA claims the reason for no stars being present in the pictures is because their cameras weren't set for the proper exposure settings, yet all the pictures have fantastic clarity. Let's say for the sake of argument on the first mission to the Moon the astronauts made the mistake of not focusing the cameras correctly. At a cost of over 24 billion dollars, NASA would not have let the astronauts on the other six missions to the Moon make that same mistake. They had several years to ensure the following astronauts were trained properly and corrected the problem.

Sorry but NASA's claims of no stars being able to show up in any pictures is an insult to the American taxpayers! NASA managed to teach all astronauts how to change film and swap batteries and filters in pressurized suits with limited mobility, but not one of the astronauts ever figured out how to focus the camera. How can anyone possibly believe the astronauts could not focus the camera when NASA displays hundreds of photos of the astronauts using the cameras before their alleged trips to the Moon? Something does not add up here!

Most of the Moon landing pictures look so perfect they would have taken a highly skilled professional photographer. But NASA asked everyone to believe the astronauts managed to repeatedly take perfect pictures, when the cameras were not equipped with a viewfinder. That's right, NASA expects everyone to believe that all twelve of the astronauts who stepped foot on the Moon were able to adjust and focus perfectly, except when it came to taking pictures of the stars. One would expect pictures taken from the Moon surface to be filled with stars as shown in this planetary painting.

Planetary Painting

What About The Far Ultraviolet Camera and Spectrograph used on the moon.

One major problem with NASA's claim was that the camera equipment the astronauts brought to the moon was incapable of taking pictures of stars. The camera they brought to the moon was the Far Ultraviolet Camera. This was invented by George Carruthers. This is the camera the Apollo 16 Astronauts claimed to have used to take pictures of the star consolations, like the Milky Way, they claimed to have seen from the moon's surface.

This Far Ultraviolet Camera was so massive and powerful it weighed over 50 lbs and had to be placed on a tripod to support. The Far Ultraviolet Camera was designed to receive dim-light stars, earthshine, and UV photographs. It was so powerful that it would have picked up the Milky Way along with other star clusters and even infrared lighting. The next two images are of the Far Ultraviolet Camera and Spectrograph. The astronaut training picture, on the left, was taken at the Kennedy Space Center, and the Moon landing picture, on the right, was also taken at the Kennedy Space Center.

NASA image: S72-19739 02
September 1972

NASA PHOTO ID# 10075848 Apollo 16 Mission-1972

George Carruther's invention of the Far Ultraviolet Camera and Spectrographic is truly a scientific masterpiece, which has been used frequently by the space shuttle program to make amazing discoveries. However, the way NASA has distorted the evidence produced from this camera for their Moon landing hoax is a disgrace.

This camera used in the Apollo 16 Mission supposedly produced about 200 photos from the moon surface, so where are they? According to this camera's design specifications, it would have revealed many new deep-space objects from the perspective of the lunar surface. Some of these objects were far ultraviolet images of stars, nebulas, and galaxies as well as new features of Earth's far-outer atmosphere views of the Moon. Yet, NASA only released a handful of these types of ultraviolet images to the public and those are riddled with indications of forgery. For example, new evidence, produced by the same type of Far Ultraviolet Camera and Spectrographic, from the Earth, the Skylab and the Space Shuttle totally contradicts NASA's original Apollo 16 Moon Landing illustrations.

Below are two color enhancement images of earth; the Apollo 16 Astronauts supposedly took these with the Far Ultraviolet Camera from the Moon. On the far right, is a more recent image taken by the same style camera. It was taken in October of 1999. This camera is called the Total Ozone Mapping Spectrometer (TOMS) Earth Probe. Granted the big shaded area, above the northern magnetic axis, would not have been there since that represents the ozone hole NASA and the American Government was just beginning to create. However, it's obvious the alleged Apollo 16 images were significantly different and a further investigation of these NASA Apollo 16 Far Ultraviolet Camera images were needed.

Far Ultraviolet Camera Images of Earth

NASA 1970's Claim Of Far Ultraviolet Camera Image

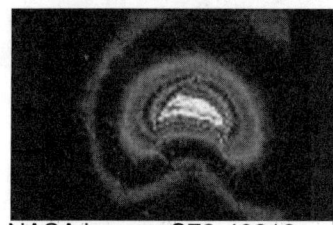
NASA image: S72-40818

NASA image: S72-40820

Recent Far Ultraviolet

NASA Apollo Moon Landing Galaxy Images Are Fakes

Using Recent Far Ultraviolet Camera To Prove the Apollo Missions Were Faked.

Below are three different photographs taken of the same Large Magellanic Clouds by Far Ultraviolet Cameras. The one on the left, NASA claimed was taken by the Apollo 16 Astronauts from the moon's surface, and the other was taken several years later on the skyLab. Two are genuine and the other is not. By now, without even looking at the photographs, you can probably guess which one is fake. It's the one supposedly taken by the Apollo Astronauts while on the moon's surface. This galaxy photo, taken with three Ultraviolet Instruments, displayed radiation from the upper half of the ultraviolet (UV) spectrum. (Photos obtained from National Aeronautics and Space Administration Washington, D.C. 1979)

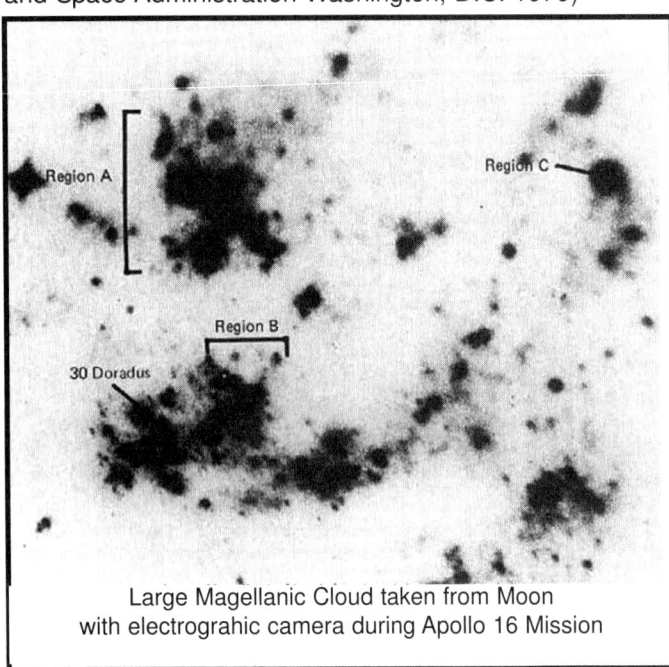

Large Magellanic Cloud taken from Moon with electrograhic camera during Apollo 16 Mission

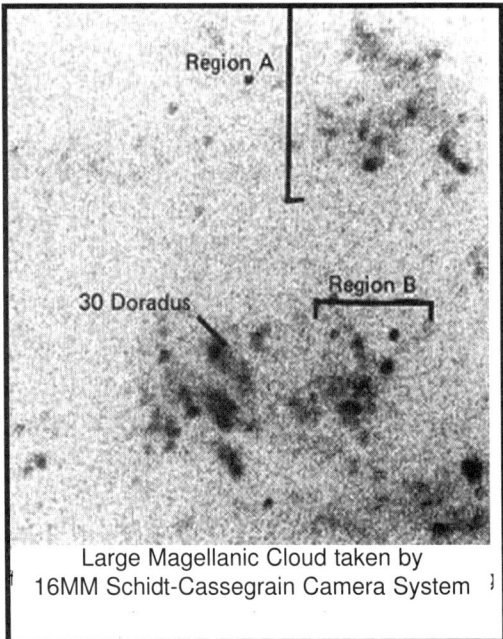

Large Magellanic Cloud taken from the Earth

Large Magellanic Cloud taken by 16MM Schidt-Cassegrain Camera System

Here is a detailed explanation of the various problems with NASA's Moon Surface Galaxies photograph compared to all others.

When comparing the alleged Apollo Moon ultraviolet photos to the ones taken on the Earth and on the Skylab, there were several significant irregularities in relationship to far ultraviolet patches. The overall structure of the large Magellanic Clouds differed considerably.

First, the Earth and Skylab photographs showed a 'bar" from which undeveloped spiral arms trailed. The alleged Moon photograph suggested that the dynamic center of the galaxy lied in region A, and that one of its major arms spiraled outward in a clockwise direction. Additionally, it split into two arms near region C. Nonetheless, in the Earth and Skylab images, region A was not the dominant force in this galaxy. It was the 30 Doradus nebulosity region.

Another indication that NASA Moon Ultraviolet photos were fakes is that region B is even slightly brighter than the galaxies focal point, 30 Doradus. Based on all the pictures taken from the Earth, this condition was incorrect in the NASA Moon photo. The Moon suggested that region B had more hot new-born stars than the complex around 30 Doradus, which was not the case in any other Ultraviolet photo of this galaxy.

Clearly, NASA's alleged Moon landing photography totally contradicts the basic structure of the galaxy. Hence, this proves their Moon landing portraits were fakes. Remember, this galaxy image would not have changed much for millions of years. Thus, NASA simply air brushed their original Earth based telescope photo of this

Could it be that NASA's engineers simply realized that any attempt to artificially fake the star constellations without being caught would have been impossible? The answer is yes. It appears NASA did indeed have some of the brightest scientific minds working on the Apollo Moon landing project. If they were to avoid getting caught, with the primitive technology available at the time, NASA's engineers decided that a solid black background was the only alternative for the Moon landing pictures.

NASA's first attempted to Fake Stars in the pictures

The following evidence supports the theory that NASA at first tried to fake the stars in the background but found it impossible. Below is a picture of a special piece of equipment developed by NASA called "The Apollo Project - Star Projector Machine" that projected star scenery onto a black background using a point-light-source projection technique.

The Star Projector works on a concept developed by an inventor named Spitz. It consists of a point-light source reflecting off a centrally located highly reflective sphere, which directs the light outward through the many holes representing the stars. The sizes of the holes are varied to compensate for star magnitude. The star images are brought into focus on the inside of the walls by lenses glued to the surface of the projector. The diameter of the projection sphere regulates the focal length required for these lenses.

NASA realized early on that it would be impossible to get the star constellations to line up to where they would not be noticed as fakes, so they elected to go without any stars in the photos. This picture was obtained from NASA's own archives. Besides it is a NASA picture of the background stage curtain that was to simulate the blackness of outer space.

The picture below is the Apollo Moon Landing Project - Star Projector Machine (secret code named Starball)

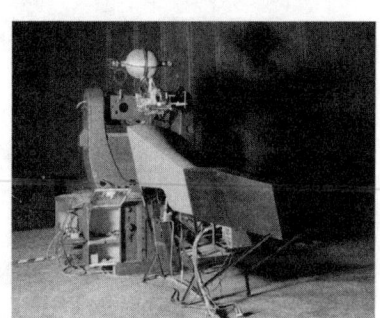
EL-2002-00439

NASA Photo #10074705
Starball testing

Below are some of the other simulators designed to be used with the Star Projector to produce the most realistic looking space travel missions, imaginable. How these additional simulators were used to fake help the fake the Moon landings will be discussed later in the book.

NASA ID# EL-2002-00337

10073602

Two Moons

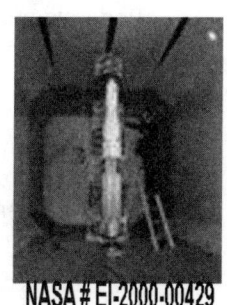
NASA # El-2000-00429

Why the American Government So Abruptly Cancelled the Apollo Moon Landing Missions

Many believe the Apollo Moon Landing Missions were canceled due to the alleged enormous cost of the NASA budget. While others believe they were cancelled as a result of a lack of interest by the American's, which was not the reason either. Since the cancelation of the Apollo Moon Landing Project, NASA spent almost a trillion dollars of the taxpayer's money.

There is a much more logical explanation as to why the American Government suddenly cancelled the Apollo Deep Space Project. This is backed by an overwhelming amount of solid evidence.

NASA Stopped Sending Men To the Moon Because Of the Stars In Space

Under pressure, to provide Apollo mission photos that contained stars, NASA made the mistake of trying to pass images of containing stars. They were immediately detected as fakes. This next photo was supposedly taken during the Apollo 17 mission, which was the last of the Apollo missions to the moon.

Unfortunately, for NASA, they made the mistake of distributing these two totally different photos of the earth, which were alleged to have been taken by the Apollo 17 Astronauts. The first one shows the earth with a completely black background, while stars surround the other.

NO Stars
as17-148-22726

STARS
including alleged Big Dipper.
as17-148-22726 a17earth11825

The releasing of these two photos simultaneously drew a great amount of attention. Anyone with some basic understanding of astrology, and some spare time on their hands, could calculate the distance with the phasing of Earth's orbiting cycle.

When this was brought to NASA's attention, they immediately cancelled all future Apollo missions to the moon. Then quickly sealed most of the information related to the Apollo project in one location protected by 24 hour armed guards, and they have not opened the room since. Surely, NASA had nothing to hide. The good news is there was enough evidence distributed, prior to being classified as top secret, to make a determination of whether or not the Moon landings were real.

The Apollo Astronaut's Own Testimony should be proof enough they never went to the Moon!

Compare the Apollo Astronauts testimonies to anyone else that has actually traveled to outer space. You will find their explanations of the experience and attitude towards the trip into outer space are completely different. The Apollo astronauts often claimed it was no big deal and nothing special and routine. Yet the people that actually made the trip into space are jumping up and down with excitement.

The first man in Space, Yuri Gagarin, a Russian cosmonaut immediately upon reaching outer space made reference to the stars being "astonishingly brilliant" during his 108-minute orbit around the Earth on April 12, 1961. Yet seven trips to the Moon and not one astronaut mentioned seeing any stars and then continuously referenced space as a vast darkness with only seeing the Earth, Moon and Sun.

When comparing the interviews of space traveler Dennis Tito to the Apollo astronauts you can see a big difference. Dennis Tito looked and talked like it was the experience of a life time and the Apollo astronauts act like it was nothing special and appear to be nervous as if they are afraid of getting caught in their lie.

Yuri Gagarin

CHAPTER 5

NASA Builds Movie Studios And Calls Them Training Simulators

EARTH

MOON

Chapter 5 -Part 1
NASA Builds Movie Studios
And Calls Them Training Simulators

From the beginning of this research, it was decided not to go public with any information proving the moon landings were faked until every part of NASA's moon landings were analyzed first. If only one clue had been released in the beginning, NASA could have easily dismissed it as just another unexplainable Moon anomaly. NASA may have even attempted to call the Earth creatures found on their moon landing film footage space aliens, in an attempt to keep their deception a secret.

Since the Internet WhizKids were collecting a great deal of evidence related to the Apollo missions, it was relatively easy to determine how NASA masterminded the entire hoax, using a simple technique I developed at an early age that I call: "The more you learn, the more you earn."

Here is how we used this technique: I had everyone collect all of the evidence together in one place, regardless of what portion of the mission it was related to. Then we spread all the evidence out on the floor. For the next two weeks I had everyone walk around, look at the evidence, and make notes of how certain pieces of the puzzle seem to fit together.

After two weeks of studying the evidence, most of the mystery was solved. At that point it was easy to visualize exactly how NASA masterminded the entire Moon Landing hoax. It was then just a matter of confirming the theory and documenting the findings in this book. Remarkably, this book now contains what many believe to be the true history behind the Apollo Moon Landing Missions, and the methods NASA used to fool 3.7 billion people, from start to finish.

The Moon Landings Were Faked In Elaborate Movie Studios

The Moon Landing Hoax all started shortly after President Kennedy made his challenge to the American people in 1961, to put a man on the moon. NASA's scientists repeatedly tried to explain to government officials that they could achieve this lofty goal by the year 1970. Frustrated, and under enormous pressure by the American people to dominate the Soviets in the space race, government officials ordered NASA's engineers and scientists to produce a realistic manned mission to the Moon that would fool the world.

NASA Movie Studio

NASA engineers and scientists went to work, creating several movie studios that would produce the most realistic moon film imaginable. Inside large buildings, NASA created lunar orbit, lunar landing, space walks, and even the lunar surface simulators; (so realistic they could easily convince anyone the Moon landings were real).

This chapter outlines precisely how and where NASA built their movie simulators to stage the six Apollo Moon Landing Missions here on Earth. This chapter contains startling, never-before-seen photos NASA claims were taken on the moon, but actually match the scenery of photos taken here on Earth during NASA training missions. The information contained in this chapter explains how NASA's top-secret lunar training simulators actually functioned as elaborate movie studios.

The location of the movie studios were only revealed after uncovering and analyzing numerous mislabeled, unedited, behind-the-scenes pictures and video footage from NASA's archives. A great deal of the film footage show the Apollo astronauts staging part of their moon landing photography here on Earth when they were supposed to be on the moon, proving the astronauts never made it beyond Earth's orbit.

Contrary to what many believe, NASA did not film the Moon landings on the Moon, in a Hollywood studio, or at the famous Area 51. NASA's movie studios were constructed at various locations including the Kennedy Space Center, Johnson Space Center, the Langley Research Center, Flagstaff, Arizona, and other locations to maintain their secrecy. These locations had everything needed to fake certain portions of the moon landings here on Earth: fake moon backgrounds, orbit simulators, man-made craters, you name it, they had it.

Evidence indicates Langley Research Center in Hampton, Virginia, was used to film most of the live action films of the astronauts and was also used as NASA's #1 top-secret research command center for the Apollo Moon Landing hoax.

Between 1961 and 1973, the Langley Research Center seemed to have more movie equipment than all the Hollywood studios combined. So why hasn't anyone heard about these things before? The first reason is Langley Research Employees back then were sworn to secrecy. If anyone told, they could face prosecution, possibly jail, and lose their pension rights. No one would have spoken out.

The second reason is, NASA has classified most of the information about the Apollo Project and sealed it in a special room under 24-hour armed military guard. The evidence is scheduled to be declassified in the year 2026, when the original people involved in the moon-landing hoax have died from old age and no one is around to take the blame for the hoax. Why else would NASA classify all the information related to the greatest scientific achievement of mankind under armed guard where no one could study it? Well, there is no need for anyone to wait any longer, thanks to the thousands of photos and films obtained from NASA; you'll be able to see firsthand, in this book the best details of the Apollo Moon Landing Hoax.

First we'll tour the actual locations where NASA filmed the Moon landing photography here on Earth. Two of the most heavily used locations for filming the fake Moon Landings was the Kennedy Space Center and the Langley Research Center. So the remainder of this chapter will provide an introduction to the two facilities and the methods NASA was using to fake the Moon Landings. We will start with the Kennedy Space Center, located in Florida.

The Kennedy Space Center

Inside the Kennedy Space Center

Based on the next set of pictures, one might think the Apollo Astronauts were only inches away from stepping foot on the Moon.

Spaceship on stills

When examining the Apollo Training Mission photography, there were very few obvious clues revealing NASA's moon landing hoax. Because of this, it begs the question: "at NASA, who is in charge of authorizing the distribution of these images to the public?" Furthermore, that person needs to be fired from the government immediately.

After NASA admits to faking the Moon landings and shuts down their cover up project, this person may be transferred to another department of national security. Based on why this person or persons have hidden the incriminating evidence to NASA's moon landings hoax, they would probably pass out the secret codes to the country's nuclear weapon silos.

For example, take this next set of Apollo Astronaut training photos. With just a quick glance, anyone can see that the equipment this astronaut was working on has been raised 2 inches to compensate for the artificial moon surface that would be added to this scene later. However, it is not the major problem. As seen in the photo next to it, one might think the reason this training photo should not have been released is because it can be matched to Moon landing locations. This is true, but that is not the reason why this training photo should not have been released.

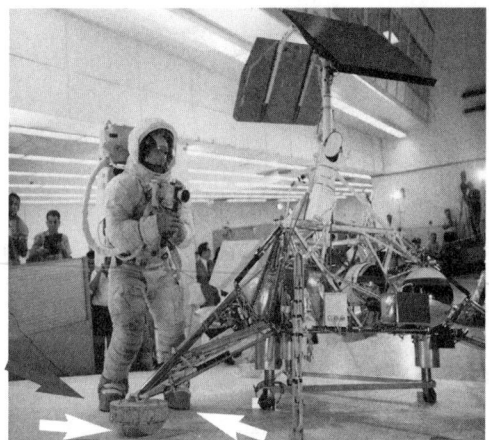
ap12-KSC-69PC-546

NASA ID# 10075419

The reason these photos were a serious problem to national security is that they revealed that the astronauts were wearing Moon boots that also had a 2-inch false bottom, which raised the astronaut. Now both the equipment and astronaut were elevated to compensate for the artificial Moon dust that will be added to the scene later.

The 2-inch false bottom, on the astronaut's boots, revealed they were specially designed for cutting and pasting the astronaut into other pictures after the actual filming took place. That is why in many pictures, which are revealed later in this book, you see the astronaut's boots were often cut off at the incorrect spot, sometimes above the 2-inch marker and sometimes below the 2-inch marker.

Why else would both the Surveyor three legs be propped up and the astronauts be wearing custom Moon Boots with the lower section of the boot extended 2 inches.

Again, this two-inch gray moon boot extension was to compensate for a thin layer of artificial Moon surface that would be added to the scene later. That way, the artist could paste the astronauts into the Moon scene and no one could tell the difference. Evidence indicated that NASA used several different techniques for creating the moon photography based on the activity. In this situation, NASA used two identical rooms for the same Apollo Mission site. One room had no moon dust with the astronauts training in it; the other room had artificial moon dust added to the movie studio. After filming the astronauts in the clean room, they simply pasted him into the picture of the same scene with the artificial moon surface.

Below are several examples showing that the astronauts had several different sets of moon boots, including one with customized false bottom for the cutting and pasting of Moon portraits.

There were many clues in these next two photos, that can prove they were taken at the some location. Later in this book, there is a whole section dedicated to these two NASA photos and the overwhelming evidence proving both were taken at the Kennedy Space Center on Earth.

Location: Kennedy Space Center   Location: Kennedy Space Center

NASA ID# ap17-KSC-72PC-379   Apollo 17 NASA ID# a17-20117331

Inside the Kennedy Space Center

Inside the Kennedy Space Center, we can see the pictures of the astronaut's training simulations also matched the astronaut's photos supposedly taken on the Moon. In the photo on the left, the astronauts were seen working beside the Apollo Spacecraft with girders shown in the background. According to NASA, this picture was taken during a soil-sampling test conducted at the Kennedy Space Center here on earth.

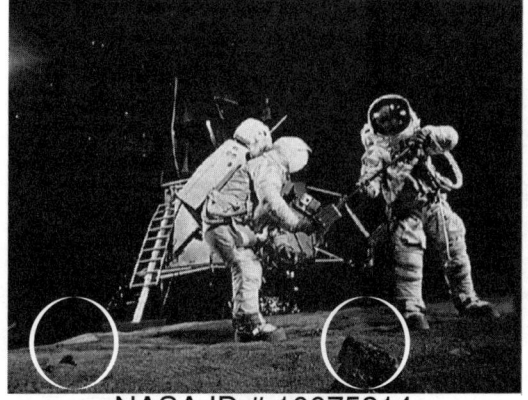

NASA ID # 10075214

The picture on the right was supposedly taken on the moon during the Apollo 11 mission, and the astronauts were conducting the same soil test. Unless NASA had a secret telescope with the accuracy to detect the exact spot the ship was going to land and find the rock arrangement, these two photos were taken at the same movie stage on earth. The large rock in the front appears in many photos and is used as a stage marker. The rock was used so often in one NASA photo it is labeled, "BIG MOLLY".

The next set of pictures show how easy it was for NASA to switch the scenery from Earth to Moon. They simply raised the black background curtains surrounding the astronauts and their work area, then dimmed the lights and started filming the astronauts, as they would appear to be on the moon. Without these pictures being available at the time, no one could have been able to really know if these photos were taken on the moon or not.

NASA - Earth Training- KSC

Earth Training

Moon - Apollo 11 Mission

NASA ID# 10075635

Black Curtain

Black Curtain

Another reason it was easy to match the training photos to the moon images was because NASA never resolved the problem of having to use high-powered stage lights for their Apollo Moon landing movies.

In these pictures, we can see the same stage light reflection is visible in the astronaut's helmet in both the Earth and Moon photos. In the picture to the far right, we can even see the curtain, from above, was lowered behind the astronauts and their ship.

NASA Builds Movie Studios And Calls Them Training Simulators

NASA's Moon surface photography shows evidence of being an artificial stage platform.

The next set of pictures is the only thing you'll ever need to prove the moon-landings were faked. Like most of the pictures presented by NASA as evidence, under close examination, it becomes obvious their pictures could not have been taken on the moon. There seems to almost always be something wrong; whether it's incorrect shadowing, excessive lighting, or something strange about the scenery, and these pictures are no exception. The next set of pictures come from NASA's archives, which they claim were taken on the moon during the Apollo 11 mission. This was supposedly the first time man had ever walked on the moon. Under close examination with a microscope, we can see the moon was already equipped with underground plumbing before man arrived.

NASA ID# as11-40-5863

NASA ID# as11-40-5864

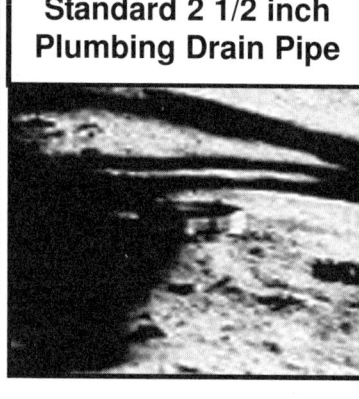

Standard 2 1/2 inch Plumbing Drain Pipe

Now of course underground plumbing on the moon is not possible, and these pictures were obviously filmed on earth and made to look like they were taken on the moon. Notice how the artificial moon dust is packed around the plumbing pipe to hide its appearance.

Secret Messages Left By Former NASA Employees

NASA claimed these next photos were taken on the Moon. These shots have a very exciting story behind them. People working on the Moon landing project, back in the 1960's and 70's, in hope that they would be found by people of the future, wrote messages in the artificial moon dust. After examining this picture under a microscope, it appeared several people even autographed their message. For more details on these messages, study the chapter called Artist's Clues.

IFrom NASA as11-40-5921

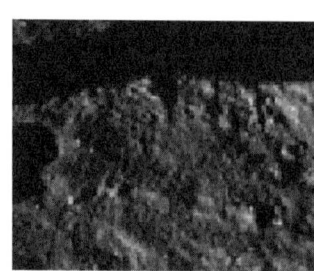

5-8

Chapter 5 - Part 2

Outside the Kennedy Space Center

Outside the Kennedy Space Center

Here, outside the Kennedy Space Center, back in the 1960's and 1970's, is where NASA conducted a great deal of the filming for their Moon Landing Hoax.

NASA claims these photos of the astronauts staging Moon Landing missions here on earth, were simply top-secret training simulations for the Apollo program, but don't believe that for one minute. These were not training simulations, the real purpose was to create realistic photographs and film, that could later be altered by NASA photographers and presented as Moon photographs for the world to see as you will see in the remaining chapters of this book.

The Apollo Moon mission training simulators are identical to a typical movie studio platform, with a relative flatness of the foreground, and a small working area for the astronauts to perform their moon experiments, usually no more than a few yards wide. This method allowed the stage crew to exchange the scenery without much difficulty. It is similar to a stadium that, within hours can be changed from a monster truck dirt pit into a ladies beauty contest. That is how they changed the scenery, it was that easy.

Oh what a lie, back in the summer of 69, oh what a lie, oh what a lie!

Construction of a portion of artificial moon's surface on the movie stage platform

This next set of pictures shows how the artificial Moon surface was applied to the Apollo Training Simulator work area at the Kennedy Space Center, which in this book is going to be referred to as the movie stage platform.

The pictures from left to right show how the surface was built up in stages. The middle pictures surface is higher than the picture on the left, and the picture on the far right has had the final layer of moon surface added to give it a very realistic look. Notice how the middle picture's surface is higher than the picture on the left. It has been filled in with the first layer of artificial moon surface. The picture to the far right shows the final layer of moon surface added to make it look very realistic.

If this movie stage platform was used to fake the Moon landings, there should have been some sign of the white sand being exposed on the Moon's surface, and there is. Not just a few grains of sand, but large amounts of white sand appeared in much of NASA's Moon Landing photography. When the thin layer of Moon dust was removed by the astronaut's movements, while working in a landing site area, a white layer of soil was exposed under a thin top layer of gray Moon dust. This made no sense. How can the Moon surface, over 4.5 billion years old, only have a thin layer of moon dust covering a white sandy surface?

The Discovery Of White Sand On the Moon

Below are three very special pictures to help prove there was white sand on the Moon and where it came from. The picture on the left was taken at the White Sands Apollo Astronauts Training Facility in New Mexico. The middle picture was taken at KSC during the Apollo mission training, and the picture on the right was supposedly taken while the astronauts were digging on the moon's surface. While the astronaut was digging, he accidentally started pulling up the movie stage platform's white sand base just a few inches below the Moon's surface.

If the American Moon landings were real, evidence of white sand just below the Moon's surface would defy science. Whether the astronauts went to the Moon or not, one thing is for sure; they certainly faked the Astronaut's Moon landing photography.

Earth's rock formations match those claimed to have been found on the moon

In the next set of photos, the two pictures on the left were taken at the Kennedy Space Center. However, those random space rocks have the same layout as the rocks shown in the far right picture that NASA claims is a picture from the Moon. Now is that strange or what, anyone can see the two pictures on the right were filmed at the same location just at slightly different angles. Notice how the larger rocks are placed in front of the smaller rocks in both pictures. It would be impossible to calculate the odds of NASA creating an identical boulder landscape before ever visiting the Moon.

NASA ID# 03fbc5e0   NASA ID# 10075925   NASA ID# 10075966

Many of NASA's outdoor filming and still photos for the Moon missions were taken at the Kennedy Space Center. The large boulders and backgrounds were simply edited for the Moon landing photos as you will see later in the book. After careful examination of the Apollo training simulator's functionality, it's clear they were actually designed to function as movie studios.

Chapter 5 - Part 3

The
Langley Research Center

NASA Builds Movie Studios And Calls Them Training Simulators

The Langley Research Center

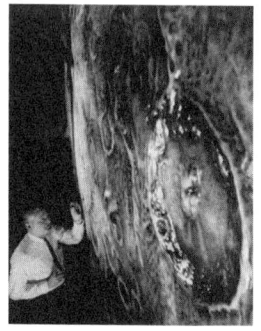

Here is NASA's Langley Research Center, which was another location where the Moon landing missions were filmed. These pictures were taken during the Apollo Project back in the 1960s and 1970s.

INSIDE THE LANGLEY RESEARCH CENTER

LOLA; The Lunar Orbit and Landing Approach simulator

Inside the Langley Research Center is where NASA simulated the space trips and Moon orbiting film footage. Here you can see how the Lunar approach and close up orbit of the Moon's surface were faked. These large Moon surfaces, shown below, were used to simulate the film supposedly taken from the Apollo Command Module as it approached the Moon, orbited the Moon, and then descended to the Moon's surface. It was hidden from the public under the code name "Project LOLA", for Lunar Orbit and Landing Approach simulator. During one of the Apollo 11 moon shots, Neil Armstrong's first sentence is "LOLA incognita" referring to "The Unknown Moon". His last sentence starts, "LOLA is once again isolated." What else could Neil Armstrong have been referring to other than this movie simulator where he was staging the moon landing called "LOLA"? Or are we to believe Neil Armstrong just happened to make up some word that just matches the word of a secret moon landing simulator only discovered many years later.

The "LOLA" Simulator

Lunar Flight Simulator

Two Moons Finished

Astronauts Film Two Moons
NASA ID# EL-1996-00199

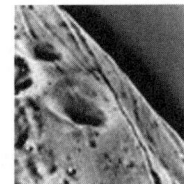
2nd Moon Reflection
EL-2001-00439

The picture on the left shows the 20' diameter rotating Moon sphere, and closed circuit TV, mounted on the guidance track. This picture was taken at the early construction stages of a special Apollo moon landing Simulator. The second picture shows the two Moon's spheres several years later with the real looking Moon surface, and craters almost completed. This movie studio is where the orbiting pictures were created for the Apollo missions. This sphere had a powerful light inside it, which was translucent on the outside. Turn off all other lighting, and you end up with a realistic Moon in the void of space, as shown in the picture on the right. No one could tell the difference, except for the lack of stars.

Notice the attention to detail as artists used paintbrushes and airbrushes to put the finishing touches on their new moon surface creations. Notice how dark the background is compared to the original picture taken at the beginning. This is because they have spray-painted the studio wall and ceiling black in preparation for the upcoming Apollo mission movie.

It's almost unimaginable the time, trouble, and expense that NASA went through in creating the authentic Moon surfaces for the Apollo movies. Here you can see some engineers checking if the artists have built the plaster craters exactly to scale and layout, as the craters are shown to be on the lunar photographs previously taken by a high magnification telescope.

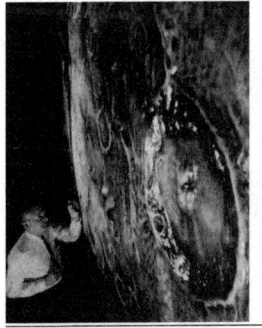

This LOLA simulator was designed to project a detailed visual encounter with the Moon surface.

The simulator consisted of a machine representing a cockpit that ran along the track and the cockpit included a closed circuit TV to record the orbiting positions. As the Lunar module cockpit traveled down the track it passed four large murals, or "scale models", representing portions of the lunar surface seen from various altitudes. With a pilot in the cockpit moving along the track, that would create the visual of being in a certain vicinity of the Moon. James Hansen, a former NASA employee, is believed to be one of the gentlemen in this picture orbiting the Moon. In records obtained from NASA James once wrote "He felt it was quite aesthetic (visually realist)".

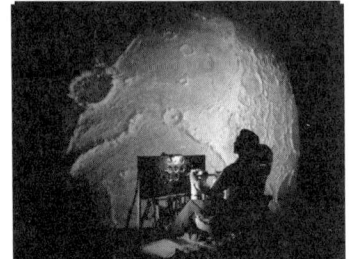
Module 1: The 20-feet sphere

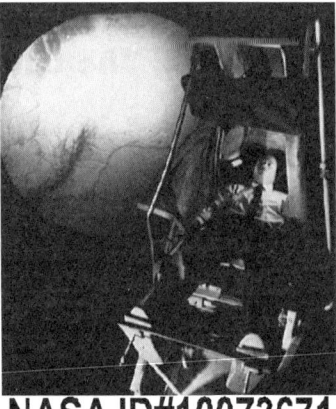
NASA ID#10073671

Here is How NASA Used The LOLA To Stage the Moon's Orbit.

After reviewing the construction of the Apollo Mission LOLA simulator and reviewing the related NASA detailed reports, it was possible to determine precisely how the Simulator worked. It was capable of producing the illusion of a spacecraft traveling to and from the moon, orbiting the moon surface, as well as moon landings and take-offs.

Two Moons EL-1996-00199

Simulation of Space Travel to the Moon: First the film makers would focus the camera mounted to the guidance track, in the direction of the Moon sphere on the right. Next, they would slowly move the camera down the track toward the Moon sphere on the right, simulating the Lunar Approach. When the cockpit reached a predetermined distance, it would stop and the Moon sphere on the right would start to spin.

Simulation of the Lunar Orbit: The small sphere spinning on the right would then create the appearance of the Apollo astronauts orbiting the moon at approximately 200 feet above the Lunar surface.

Simulation of the Lunar Landing: Here is how it could simulate the astronaut's descent on to the Lunar surface. The film makers simply flipped the cockpit camera over to focus on the Larger Moon sphere on the left. They gradually moved the camera down the track as they began zooming in on the camera toward the sphere surface.

Simulation of Return Trip to Earth: NASA film makers simply completed the above steps in reverse order.

Unfortunately, NASA took apart the LOLA simulators right after the end of the Apollo moon landing missions to cover up their hoax. Maybe destroying these simulators is the number one reason NASA has not been able to send men back to the moon!

The good news is, there was enough evidence to determine exactly how this particular simulator would have functioned back in 1969. The discovery of this LOLA simulator answered many of the questions about their alleged Moon landings that up until now have been dismissed simply as unexplainable Moon anomalies. For instance, the LOLA simulator explains how the Apollo mission films have an incorrect number of orbits completed during time frames associated with the Apollo missions. It also explains how the astronauts could have orbited the moon so quickly; occasionally much faster than would ever be possible. It was a direct result of NASA engineers incorrectly regulating the speed at which the 20-foot sphere rotated.

This also explains why the film footage allegedly taken by Apollo 8 as it supposedly circled the Moon is the same film used for the Apollo 11 mission, except that film is reversed and run backwards. Clearly No Spaceship circling the moon could have ever followed the same projectory track. NASA simply filmed the lunar surface traveling in one direction on the LOLA Simulator, then reversed the camera, and filmed the surface traveling in the opposite direction, as shown in certain videos.

It now seems obvious why astronaut James Lovell, referred to the Moon as being "essentially gray, no color, looks like plaster of Paris." He must have been referring to the oversized plaster moon surface constructed at the Langley research center.

The LOLA Simulator would also answer the question why pictures of Earth, supposedly taken from the Moon, clearly show the Earth much too small as it relates to the Moon's horizon and why the Earth's Moon Phases are inconsistent. These Earth/ Moon Phases are discussed in greater detail in the Advanced Theories Chapter of this book.

The LOLA Simulator was clearly a technical marvel for its time and certainly did not come without a great deal of trial and error. There is all sorts of evidence that proves NASA attempted several other methods of creating a realistic looking simulator with no success. Below are a few examples of some of NASA's failed attempts.

The Paper Earth and Moon Simulators: The picture on the right reveals the problems which NASA experienced, with first using a paper Earth orbiting simulator. The Earth's outer paper shell would peel off under the force of the powerful fans required to keep the astronauts cool, while they were in their space suits.

NASA ID# 10073893
Simulator tear in earth.

The Multi-exposure Simulators:

Another moon landing simulator called the Translation and Docking Simulator was constructed. It was designed to take close up multi-exposure views of space flights and space walks. Both these simulators are discussed in more detail in the upcoming chapter that covers the space walks.

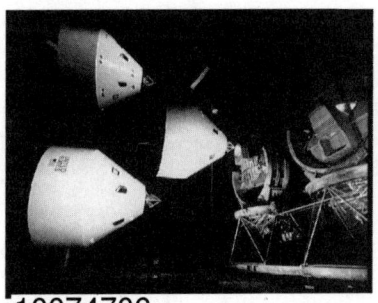
10074706

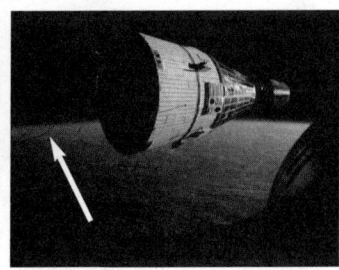
10074250

Chapter 5 - Part 4
"Moon Simulator"
OUTSIDE THE LANGLEY RESEARCH CENTER

The Apollo Moon Landing Area.
What in the world was the American Government thinking, of course they would be caught?

Let's first take a look at what was going on outside the NASA Langley Research Center.

Again, the photos used for this next portion of the moon hoax theory were all obtained from NASA's archives and at the time were available to the public. Why has no one ever questioned them is anyone's guess!

The Moon Mission Simulator

During the Apollo Moon landing era, a huge overhead crane structure was built outside the Langley Research Center. This crane structure was about 250 feet tall and 400 feet long. This simulator was capable of producing realistic film footage for many parts of the missions, including very realistic lunar landings. The Moon Mission Simulator was primarily used for the final 200 feet descent to the Moon's surface. That final descent photography would then be combined with the LOLA Simulator film to produce the very realistic Moon film footage. In order to simulate weightlessness during some of the live action moonwalks, the simulator was also used to suspend both astronauts.

LUNAR Research Vehicle

NASA ID# ap-crane

Below this huge eight-legged, red-and-white gantry (support) structure is a lunar-surface simulation. According to NASA records, Donald Hewes was working for NASA, and he was in charge of making the moon landings seem authentic. Hewes had his crew fill the base with dirt and modeled it to resemble the moon's surface. They installed floodlights at the proper angles to simulate lunar light, and installed a black screen at the far end of the structure to produce a black Lunar sky. Hewes claimed he personally climbed into the fake craters with cans of everyday black enamel and sprayed them so the shadows resembled what you would expect to see during a moon landing. Hewes took pride in creating realism.

This crane was designed to simulate the Moon landing to the point of even matching the lunar gravitational force through a servo-driven vertical cable system that supported five-sixths of the vehicle's weight. The remaining one-sixth of the vehicle weight pulls the vehicle downward, simulating the moon's gravity. A pulley system on the vehicle permits angular freedom for maneuvering the LM spacecraft around with ease.

The outdoor filming answers many questions regarding the authenticity of NASA's alleged Apollo Moon landing film. This explains the reason why the astronauts could not control the American flag from constantly blowing around in the wind. This explains why many animals have been found in the NASA pictures alleged to have been taken from the Moon. It also explains why there was never a rocket blast crater under the Lunar Module on the moon's surface; the spacecraft didn't work under its own power, it was lowered from the crane above.

Still not convinced? Try explaining how the astronauts controlled the Lunar Module spacecraft as shown in the Apollo training and mission films. The alleged spacecraft was nothing more than a chair attached to a rocket engine. The only way the astronauts could have accomplished that feat is if it were controlled by a crane system identical to this one.

Dryden Flight Research Center ECN-453 Photographed 1964
Joe Walker pilots the Lunar Landing Research Vehicle (LLRV). NASA photo

Below are several NASA photos showing the Lunar Module being maneuvered onto the crane. The picture on the left shows the stage crew in preparation for the filming, spraying down the movie studio with water. The water was necessary to keep the artificial Moon dust under control, which can be seen in this picture standing about 10 feet behind the Apollo Spacecraft. That is the portion of the stage where the actual filming of the moon landings took place. With this much Moon dust on the movie studio, if the surface was not moistened, the camera lens would have quickly become contaminated.

The two pictures on the right show how the Apollo LEM Spacecraft, with astronauts inside, were then positioned over the artificial Moon surface to begin filming the final stage of their descent on to the Moon's surface.

Okay, lets for a minute imagine the Apollo Moon landings actually took place. Then why would NASA be toying with Moon landing movie simulators that completely match detailed lunar landscapes at the same time the astronauts were already traveling to and from the Moon? Maybe they were faking the Moon landings all along?

The pictures were supposedly taken around the time the first Moon landing, in 1969, was to have taken place. The first landing consisted of astronauts Neil Armstrong and Buzz Aldrin. The picture on the far right is known as a Multi-Exposure time-sequence of the Lunar Module traveling from one end of the crane to the other, while simultaneously being lowered. Hence, an authentic looking Moon landing was created.

Notice the lights that were fixed at the top of the crane structure; seem to match the ones in the Moon pictures. There also appeared to be a large fan under the spacecraft, which could have created the dust scattering effect of a rocket engine as it descended on to the fake Moon surface.

The film showing the LM Spacecraft, supposedly blasting off from the Moon's surface, was also created beneath this crane at the Langley Research Center. The Lunar Module could have simply been attached to the crane, and hoisted rapidly. Furthermore, while it was hoisted, the blast off sparks were ignited beneath it.

The Langley Lunar Crane Was Also Used To Simulate Certain Portions Of the Moonwalks.

NASA's filmmakers had several methods for simulating the low gravity environment of the Moon. For the action scene on the Moon surface, the astronauts would often be suspended from the Langley Lunar Simulator Crane by several strategically placed elastic bungee cables. This can be seen in the picture to the right. Throughout NASA's Moon Landing Film footage, there was plenty of evidence revealing this gravity crane system was used to produce the Moon walking activities. This evidence will be covered in great detail later in the book. So there you have it. You've seen how the Lunar approach, orbit, landing, and take-off were accomplished. So the next time you see a film on TV of any Apollo Spacecraft supposedly approaching, orbiting, landing on, or any astronauts taking a walk on the Moon, you will know exactly how it was done.

CHAPTER 6

How NASA Simulated Weightlessness of Space

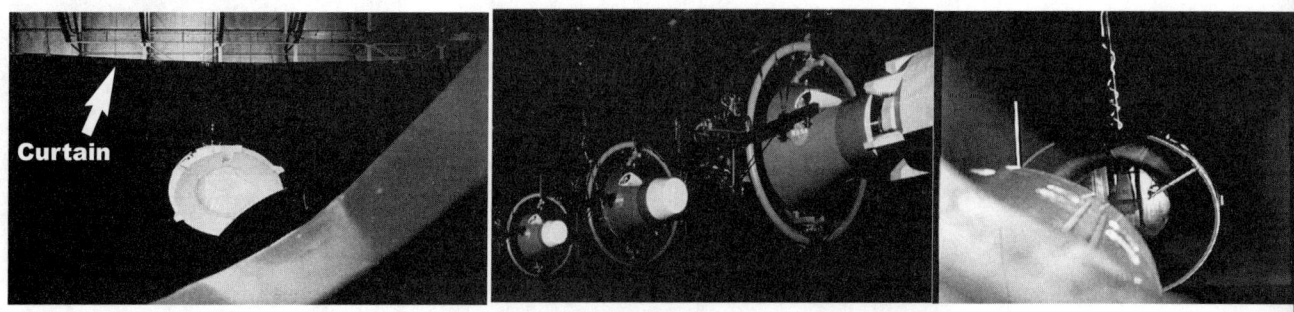

It's Showtime

Chapter 6 - Part 1
How NASA Simulated the Weightlessness of Space

Astronaut Training Simulators or Movie Studios, You Decide

Analyzed under a microscope, every piece of NASA's Moon landing evidence indicated it was just a 56 billion dollar movie. This next set of training simulator images most defiantly reinforce the movie theory. To make it look as if it were filmed in outer space, these snapshots reveal how NASA used artists to remove objects from the training simulator photography. Take for example, these next three photos obtained from NASA. They illustrate how easy it was for NASA's artists to convert the astronaut training photos into the most realistic looking space film footage ever imaginable. These three photos are truly an amazing discovery. It's a wonder why NASA ever released them to the public. These photos demonstrate the step-by-step process NASA's filmmakers used to convert the training film into images of outer space.

The first picture displays the trolley and cable system supporting the ship in a simulated space environment. The middle shot shows a multi-exposure sequence with the support trolley and cables painted. The third training simulator shows the two ships in outer space with planet Earth below it.

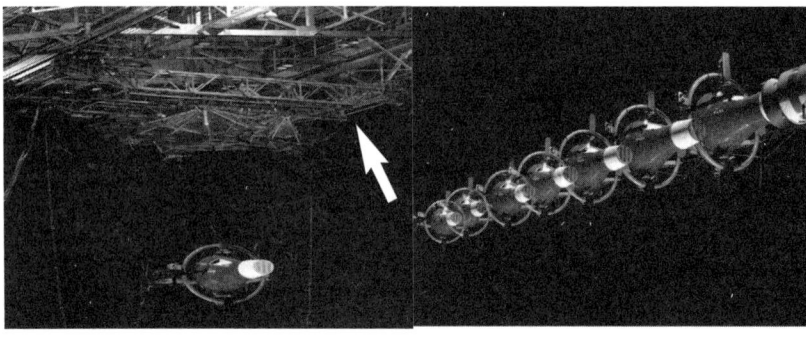

NASA ID# EL-1999-00386   NASA ID# EL-1999-00385   NASA ID# EL-2001-0039

In order for the middle photo to look this way, it would have required at least seven individual pictures of this ship as it moved along the above guided track. Furthermore, the picture to the far right shows the Earth painting is added under the two spaceships. Once again, proving NASA produced film footage here on Earth that looks identical to what would have been seen in outer space.

The sample below shows how easy it was for NASA to convert the training simulator photography into an authentic looking space film footage. The picture on the left is the original NASA training photo. Beside it, is a copy of the image showing that the ship levitation harnesses were covered up. Take a close look at the astronauts climbing out of the ship without his space suit on. Does it look like he is about to jump back down to Earth, or is he in a training simulator.

NASA Training Simulator photo

Sample covering harness bracket with just the color black and reducing brightness by -15.

Many of NASA's images were so informative, it was hard to imagine they ever slipped by NASA's screeners and were released to the public. For instance, a large variety of earth creatures can also be found in many of the photos NASA claims were taken in outer space and on the Moon.

With something so contradictory as animals in outer space, the question was not if NASA faked the Moon landing events, but rather how they devised the event. After uncovering what methods NASA had been using to fake their space flights back in the 1960's and reviewing all of the available Apollo Spacewalk photography, how and why the Moon landing images exhibit illustrations of animals? NASA's claims of the photography being taken in outer space are simply not true.

Most of the outer space flights were filmed in large customized airport hangers (warehouse buildings). That is why birds can be found in the space flight pictures. They would often fly into the airport hangers and land on the spacecrafts. Since most of the Moon surface activities were filmed outdoors, that is why we see many animals in NASA's Moon landing photography. It really was that simple!

Birds Hitch A Ride To Moon On Spacecraft.

These birds simply flew into the airport hanger where the Agena Target Docking Vehicle from Gemini 8 Spacecraft space missions were being faked. They landed on the spacecraft and were detected by the photography editor. NASA then released this photo many years later in error.

Black Bird

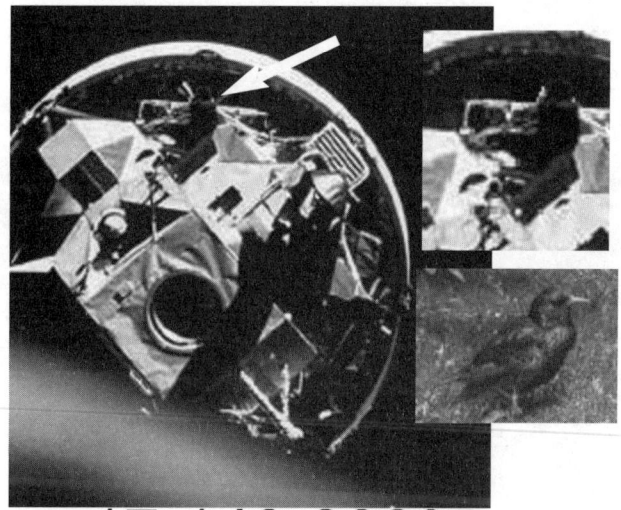

as17-148-22695

Chickadee Bird

NASA ID# 10074212

A small Chickadee Bird

An animal pops it's head out of an astronaut's open helmet

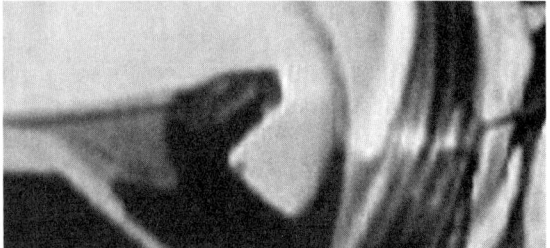

A snake peaks out of a helmet

Reflection on helmet shows a bird flying by

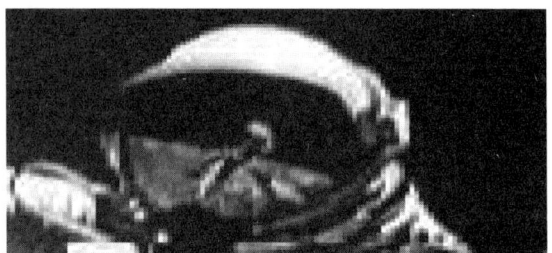

Bird Flying by in Moons Orbit

A small animal rests on Astronaut

NASA ID# 10074015
Bird and or small animal on astonaut dummy

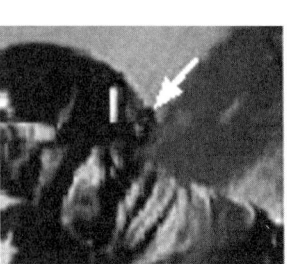

NASA Also Took Advantage Of Physics To Produce Weightlessness In A Free-falling Plane.

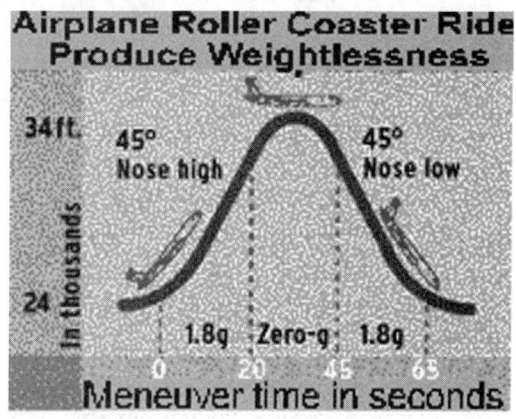

For many space-walk scenes, NASA had the astronauts act out the space walks in a special customized Air Force KC-135 plane. This plane was nicknamed "Vomit Comet". (Comparable to a roller coaster, it had a tendency to upset people's stomachs). The astronauts and filmmakers got into this plane and climbed from 24,000 to 34,000 feet at a 45 degree angle. After reaching that height, the aircraft leveled off momentarily. Once again, the flight felt like the top of a roller coaster. Next, the airplane tips its nose down 45 degrees and accelerates downward. To counteract the drag of air resistance, the pilot guns the engines. As a result, there is about 25 seconds of weightlessness. As the plane climbs up again, the cycle of zero gravity followed by double gravity is repeated.

The projectory of the plane makes the passengers seem like they're floating in space, when in reality, they are falling like a rock toward the ground.

Space Weightlessness In "Airplane Free Fall" Matches Some of the earlier Gemini Missions and Apollo Outer Space Scenes.

As seen in the pictures below, using the described free-fall techniques, the images obtained from NASA show the Astronauts simulating weightlessness in a specially customized aircraft.

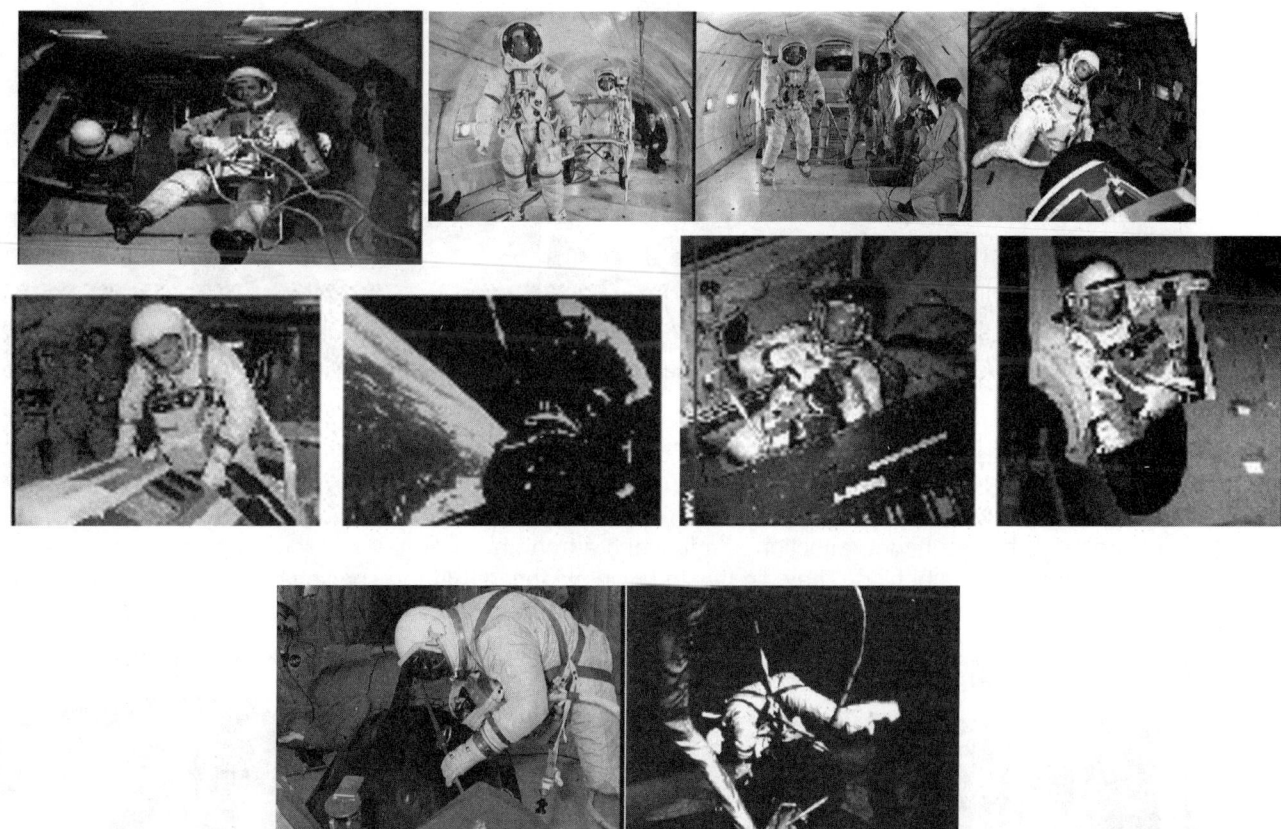

Although NASA's reduced gravity program, "Vomit Comet", was effective in producing a very realistic looking space film footage, the process was extremely hard on the astronauts and crewmembers physically. Every time the zero gravity condition stopped, the astronauts and crewmembers would crash down onto the plane's floor. People on the air force base must have known when they saw someone wearing one of these Weightless Wonder Crew Patches that the person was assigned a very hard job.

The Neutral Buoyancy (Float) Water Tank Simulator

NASA needed to come up with a better system that was easier on the Apollo Astronaut's bodies and that could extend the zero gravity time frame. With a maximum of approximately 30 seconds in the Vomit Comet plane, it required too much work to film any pro-longed space walking film footage.

In 1968, shortly before the first alleged mission to the Moon, NASA discovered water was the answer to all of their problems and quickly constructed the first Neutral Buoyancy Simulator at the Marshall Space Flight Center.

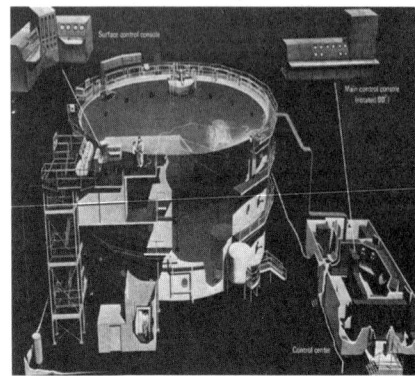

Neutral Buoyancy Simulator

By attaching a system of floats and lead weights to people and objects, engineers countered the effects of water on NBS test subjects to closely simulate weightlessness. The floats and weights compensated for water displaced by a subject's volume and density. Thus, making the subject "neutrally buoyant" -- it neither sank nor floated.

This Neutral Buoyancy Simulator (NBS) micro gravity simulator was far superior to the Vomit Comet and much safer for the astronauts. Since the Neutral Buoyancy Simulator (NBS) could produce the most realistic looking space walk film footage ever imaginable, filming for the Apollo Moon landing movies switched to this location almost immediately.

These next set of photos show how the (NBS) produced the film footage used in the Apollo space walk scenes.

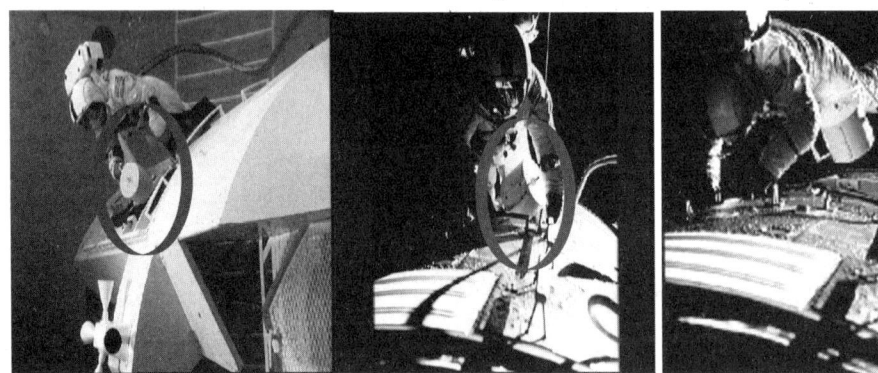
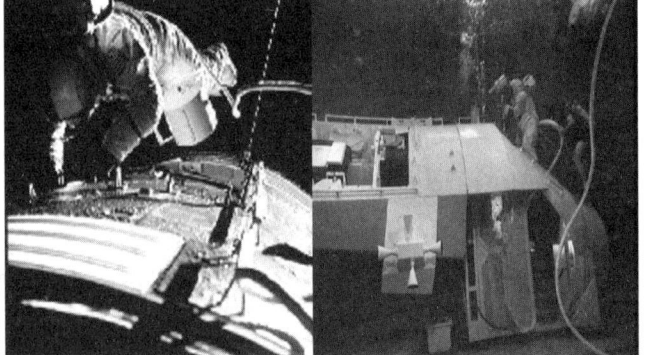

NASA's Airbrush Artists Left Clues That the Space Walks Were Actually Filmed Underwater.

An astronaut's helmet had a reflection of a fish and a frog on his arm. While touching up the photography, the artists left these clues behind. They did this to leave evidence that the spacewalks were filmed underwater in one of NASA's Neutral Buoyancy Simulators.

Space Environment Simulation Laboratory:

This Space flight simulator was stored in Building 32 at the Manned Spacecraft Center (later renamed Johnson Space Center). It contains two monstrous vacuum chambers that can simulate the conditions of space except for weightlessness. The entire spacecraft was placed in these chambers. Additionally, the spacecraft was subjected to about one ten-millionth atmospheric pressure (equivalent to 130 miles altitude) and temperature extremes of -193°C to +127°C.

The entire space flight missions could have been simulated inside these chambers and not a soul would have been able to tell the difference. In 1968, astronauts Joe Engle, Vance Brand, and Joe Kerwin spent a week (June 16-24) simulating a space flight mission.

obtained from NASA website at www.hq.nasa.gov/office/pao/History/SP-350/i4-

How NASA Simulated the Weightlessness of Space

Apollo Lunar Module Mission Simulator. Photo ID: 10074712   Date Taken: 01/11/67

This simulator was located in building 5 at the Johnson Space Center and was only one of several simulators NASA had constructed. These types of simulators were designed to give the illusion that the astronauts were traveling through outer space. From inside the capsule, photographers filmed astronaut activities. As seen in this NASA picture, each capsule window was covered with a camera. Furthermore, this camera could project realistic looking space travel films. The movies were produced on the LOLA Simulator, which was described earlier. Because of special cameras and simulators, astronauts appeared to be traveling in space; however, they were only in a simulator.

NASA ID# 10074712

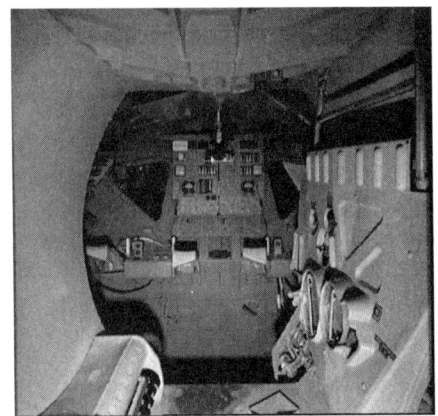
NASA ID# 10074713

While the astronauts were in these simulators, there was absolutely no way to tell if they were in space or not. The simulators were that good. To prove this point, according to NASA, each picture below was taken in a simulator. Can you tell which of these pictures were taken in space and which were not?

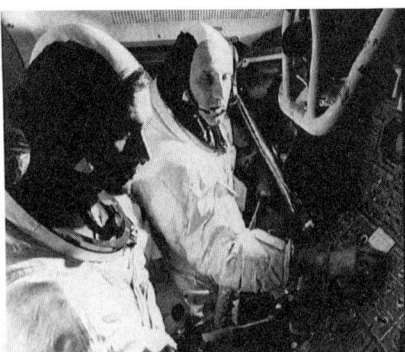
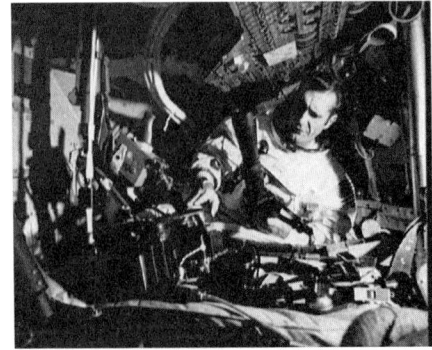

Another important point to note about the simulators is that they were built at least two years before the astronauts allegedly went to the moon.

Plan X Docking Simulator- Date Taken: 09/13/65

The Gemini 6 Spacecraft (right) and Agena Target Vehicle (left) simulate a space mission that docked on the Boresite Range Tower.

Space Simulator
NASA ID# 10074194

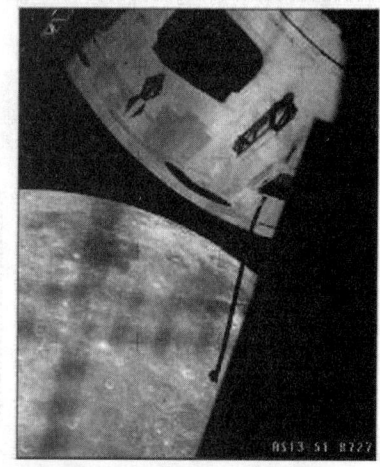

Apollo Space Image reveals was Simulator
NASA ID# 10074194

Oil Tank Simulators

The two pictures below show one of NASA's earliest attempts to produce the illusion of dark space travel. It shows just how far NASA was willing to go to make it look as if the astronauts were traveling in space.

The picture on the left is a multi-exposure view of a translation and docking simulation which was taken on August 1, 1967. At this time, astronaut Vance D. Brand was in the Lunar Module simulating a space mission.

The close-up picture on the right shows astronaut John H. Glenn Jr. exiting from the capsule during an egress (climbing out) training activity. This picture was taken at the Langley Research Center in building #5 on December 12, 1960. Mr. Glenn is attempting to transfer a life raft from the Mercury Capsule. This photo is disturbing because the spaceship is sitting in a pool of oil. Because of this, people may believe that this alleged training was most likely designed to produce deep space photography. Hence in the future, NASA could declare these as real shots.

By looking at the astronaut's hands, a person can spot this as an oil tank. Furthermore, they are covered in oil. While reaching for the raft, as John began to climb out of the capsule, he must have dipped his hands in the pool of oil the ship was immerged (placed) in.

Oil Tank Simulators

10074706 Ship above pool of oil

10073603 Astronaut in pool of oil

10074250 final results

Below are a few additional Simulators Photos that show how NASA faked the space flights:

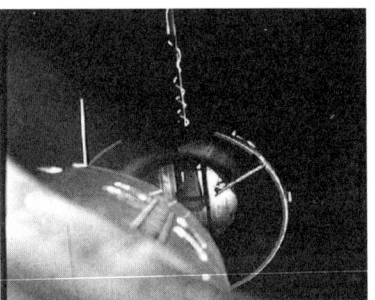

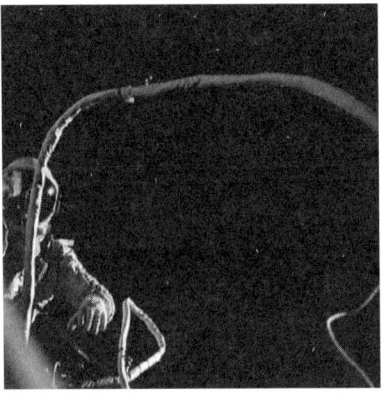

as9-20-3064

10074437

CHAPTER 6 - Part 2

How NASA Faked The Moonwalk

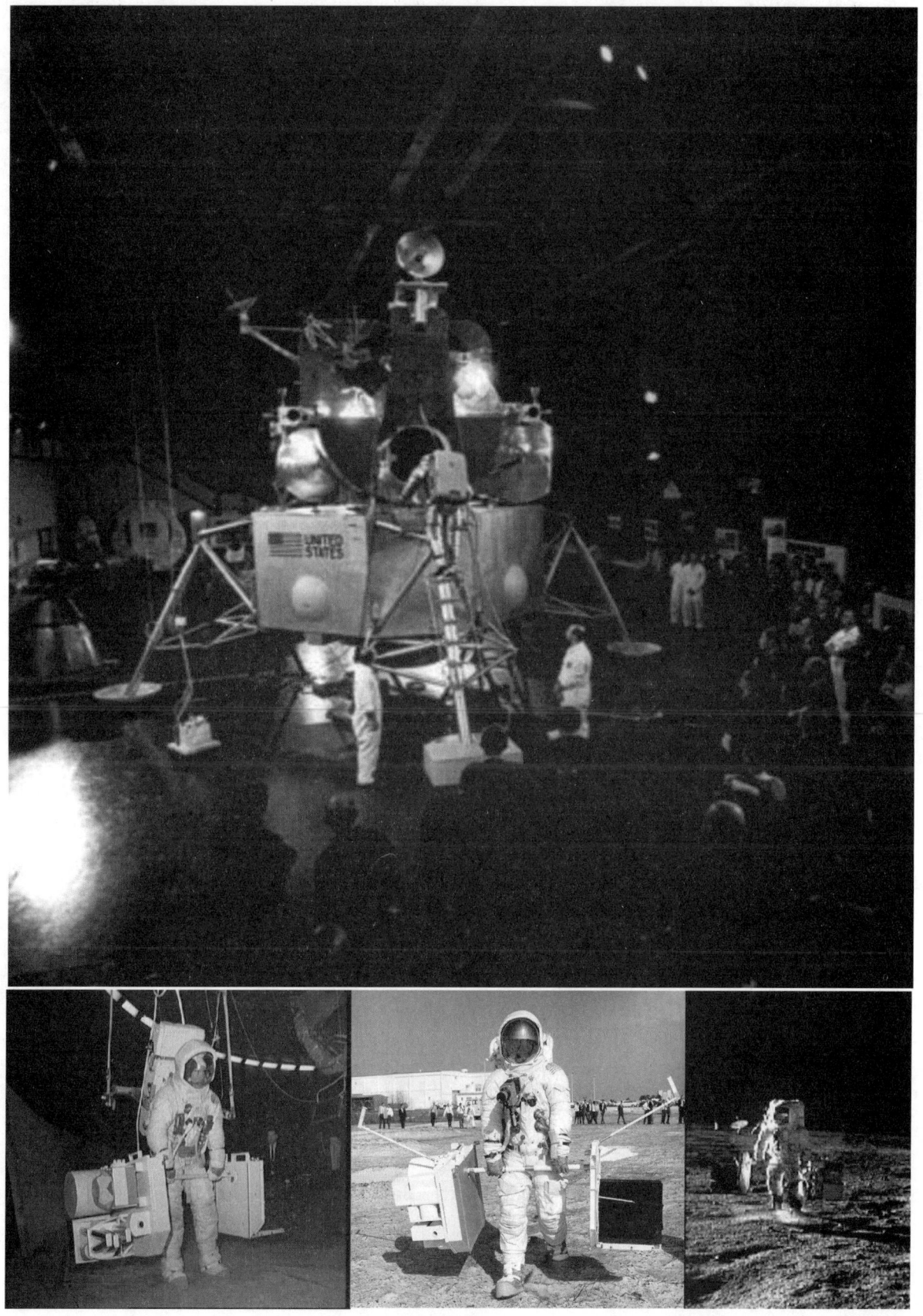

How NASA Faked The Moonwalks

If NASA faked the moon landings here on earth, many people still wonder how they simulated the weightlessness of outer space.

The answer can be found in these next three pictures supposedly taken on the moon. After having these graphics analyzed to see if they were truly taken on the moon, it was determined that they were fake and here is why.

20130785

20130786

20130788

The device that took these three photos was entirely different from the gadget used during the astronauts missions to the moon. The biggest indicator is that these photographs lacked "Crosshairs". These were distinguishing features that all Apollo mission cameras used. According to NASA officials, the + patterns were supposed to be present in every moon landing photo, yet here they are not. Again, since there are no crosshairs in any of these photos, they were not taken on the moon. If these photos are not from the moon, then where were they taken? Additionally, how did NASA make them look so realistic? A detailed explanation of the crosshair pattern is discussed later in this book.

If people believed that NASA really had such highly advanced technology to send men to the moon, then why is it difficult for many of these people to realize it would have taken much less to fake it. As we go through all the evidence, you'll see how NASA used several different methods to create the realistic looking space photography. Although they proved to be very effective, most of the images were simple and primitive. To create the spectacular simulations of weightlessness in outer space, NASA used a combination of indoor and outdoor simulators.

This next group of NASA photos shows how NASA used several different methods to simulate the space walk photographs. NASA's artists and photographers used a simple "Cut and paste technique" to create the illusion of weightlessness on the moon surface. There really is no difference between the earth astronaut training and the alleged moon landing photography. Evidence suggests weightlessness was by far one of NASA's most difficult tasks to achieve; it took much work and some real creative thinking. If they had not recognized the need to combine outdoor with indoor footage, their hoax would have been detected much sooner.

Why would NASA use fake space walk photography? Back in 1969, couldn't NASA simply have sent the astronauts to outer space, so they could take some spacewalk pictures? This may have been a possibility; however, there is no evidence that shows NASA had the technical ability to conduct these types of space walks.

Even if the Americans had tried to launch their astronauts into outer space to film the spacewalks, they still had no way to get to the Moon and conduct the alleged spacewalks there. Remember, we have already proven that humans cannot travel more than 800 kilometers (500 miles) above earth's surface. Therefore, that is the reason there were so many examples of the astronauts faking the spacewalks. They had to fake them because there was no other alternative.

Additionally, this also explained why NASA focused a great deal of resources into making the spacewalks as real as possible. If they could fake the spacewalks, it would have been the one portion of the mission that could fool everyone! Remember, back in the 1960's and 1970's, a space walk was incomprehensible to almost everyone. Therefore, when NASA held back most of the evidence from the public, other than the stuff they wanted people to see, it was impossible for an independent investigation of NASA's claims to take place. Up until now, no one has ever done what would be considered a thorough investigation of NASA's alleged Moon landings. This also explained why most people believed the astronauts' trips to the Moon were real. Moreover, they assumed the scientific community confirmed the evidence, which is clearly not the case! NASA simply escaped undergoing any type of real scrutiny by the scientific community. Well, it's about time they did. Don't you think?

How NASA Simulated the Weightlessness of Space

The Life Support System

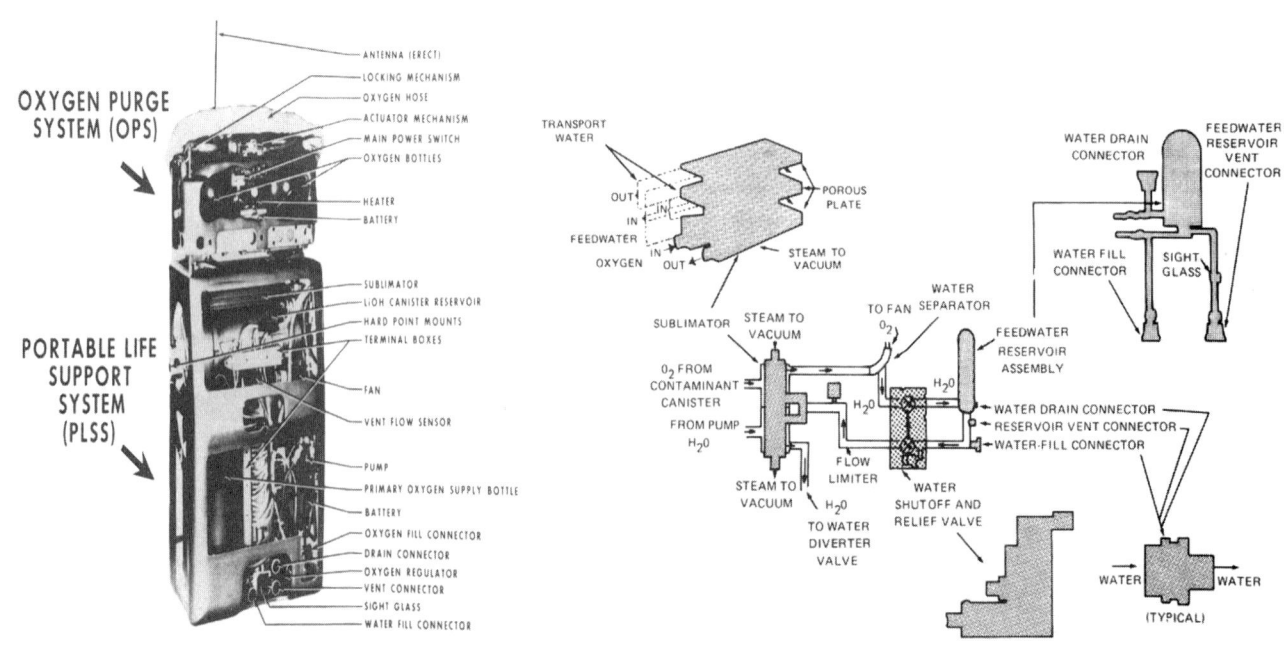

The Life Support System Was ONLY A Levitation Device Used To Help Simulate Weightlessness

Several of the young adults who worked on this research project asked about the backpack life support system. If the astronauts did not go to the moon, what was the purpose of the system? That's a really good question. Here is the answer: the astronauts backpacks were used as support harnesses to simulate weightlessness.

After obtaining the detailed specifications of the astronauts' life support system, it was obvious that the backpacks were nothing more than elaborate levitation devices. As seen in the picture below, they included side mount levitation harnesses, three dual support brackets, and a suspension reinforcement plate.

Upper Levitation Harness Bracket

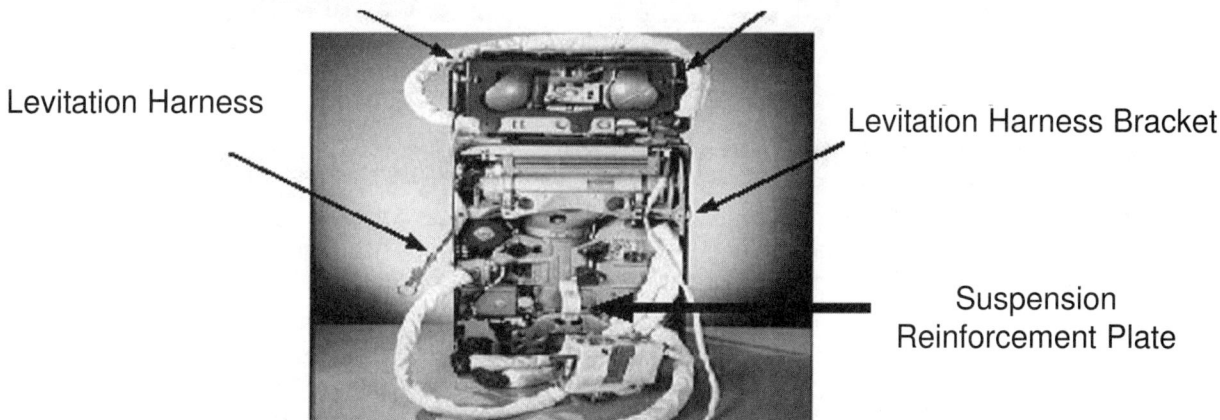

From NASA's Photo ID# 10074699
Levitation Device disguised as Life Support System

The next set of NASA Apollo training photo's shows how NASA used these custom backpack harnesses to simulate the weightlessness of outer space during the astronaut's training here on earth. These pictures also illustrate how, during the astronaut's training, the astronaut's backpacks were actually used as levitation devices to simulate the weightlessness of outer space.

NASA ID # ap11-s69-38517

NASA's filmmakers used two systems for simulating the 1/6 gravity of the Moon. EAch of the two systems served a different purpose. One was called the Horizontal Gravity Reducer, which was used for the simple moonwalk activities such as bouncing around on the Moon. The Horizontal Gravity Reducer will be explained later in this chapter. The second system was called the Vertical Gravity Reducer, which assisted in scenes requiring; climbing and lifting of heavy objects. This Vertical Gravity Reducer used the astronaut's life support system as a levitation device.

The Vertical Gravity Reducer

In these pictures, we can see how the photographers tried to camouflage the levitation brackets on the Life Support Backpack. They did this by having the astronauts switch backpacks between scenes, or simply covering the Life Support Backpacks with decorative nylon covers.

 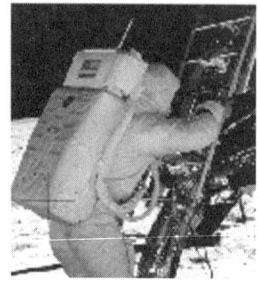 

Levitation Harnesses Found on Moon.

In this next picture, NASA wanted everyone to believe it was just a routine astronaut training. It can be proven this was actually filmed for the astronauts' upcoming alleged Moon hoax mission. This has been proven with the discovery of the same levitation bracket device found in NASA's Moon landing pictures.

In several Moon landing mission scenes, this type of levitation hook-up would have been almost impossible for the photographers to completely hide the brackets. By searching for levitation brackets, in Moon landing photography, it was possible to confirm that NASA is claiming these training films produced the film footage from the moon.. After a quick examination of the Moon landing photography, evidence of their levitation devices being used was found. The images below show one the astronaut's Earth training levitation harness used to simulate weightlessness, it is leaning up against the spacecraft alleged to have been on the Moon. NASA really should do some explaining here, but they probably won't. Therefore, we'll clarify the findings for them.

NASA ID# 10100266

NASA ID# 10100265

How NASA Simulated the Weightlessness of Space

Astronauts Found Wearing Levitation Backpack Devices On Moon

Many of NASA's portraits reveal the astronauts were wearing these levitation backpacks on the Moon. After investigating the various Moon landing photos, a remarkable discovery was made. The astronauts were switching backpacks between scenes.

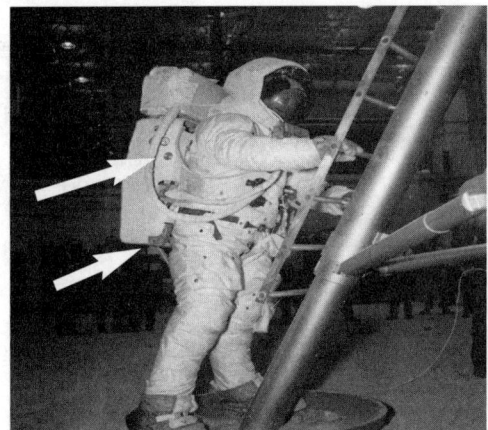

NASA ID# ap11-s69-31042

More Evidence of Levitation Backpacks Being Used on The Moon

These next sets of photos also reveal that the backpacks the astronauts were wearing on the Moon were actually the levitation bracket modules. The picture on the left is from the Apollo 15 Mission. The picture in the center is from the Apollo 16 mission, and the picture on the right is from the Apollo 17 mission.

as15-85-11514

NASA # as16-134-20425

ap17-134-20386

The levitation devices were camouflaged as a Life Support Systems that were used for these Moon landing scenes. Once again, with the evidence provided by NASA and a little common sense, one can easily discover that all the alleged Moon landings were faked here on Earth!

Astronaut With Levitation Harness Attached On Moon, "BUSTED"

NASA's touch-up artists made a very big mistake with the next moon landing photo. They did not totally edit out all of the levitation harness attached to the astronaut. That's right, one of the levitation straps can still be seen attached to the left side of the astronauts backpack, which was used to simulate gravity while he was supposedly on the Moon. This discovery confirms they were not on the moon as NASA claims and simply were being filmed in training simulators.

ap12-ksc-69pc-545

Levitation cables left attached to the astronaut Pete Conrad

Below are additional photos taken of Apollo Astronaut Conrad supposedly taken on the moon. However these images show that the levitation harnesses are not attached and are air brushed over.

as12-46-6791 Levitation Harness

AS12-48-7136 Harness Airbrush Out

With these clues to the moon landing hoax being so obvious, one has to wonder if NASA's intentionally distributed this defective Moon landing image. in hopes that it would be discovered. The reasoning is: some twenty plus years ago, this picture was brought up for debate as a questionable Moon landing photo by NASA themselves. Now, unless it revealed clues to their fake Moon landings, why would they do such a thing?

NASA Invented New Movie Special Effects Technologies to Fake Moon Landings

NASA Invented Movie Special Effects Technologies to Fake Moon Landings

As the mystery was unraveled, a lot of people began to ask if anything good came out of the Moon landing hoax. Of course, like any great adventure, some sort of new inventions were developed and the Moon landing hoax was no exception. NASA created many new technologies that filtered into public use. American's hard earned tax dollars allowed NASA's engineers and scientists to greatly advance special effects equipment for the movie industry.

In the next set of pictures, we witness NASA filmmakers using a technique heavily used in film and video special effects known as the Chroma Key (a.k.a. blue screen). The Chroma Key technology allowed the astronauts to place any object in front of fake background or a certain scene. You can watch this every night on the during a news broadcast. Chroma key technology allows a weatherperson to stand in front of an animated weather map full of flying clouds and rainstorms. Chroma key was also used to make Superman fly in "Superman" and Ewan McGregor jostle with Jar Jar Binks in "The Phantom Menace."

Using this technique, the astronauts could be seen in the following imaginary situations: climbing down a spacecraft ladder, floating in outer space, inside the Apollo Spacecraft, or jumping on the moon. Furthermore, each situation looked completely real. At present, the technique is used so often that folks don't realize it. News reports can be made to look like they are at locations when they are not. Complete segments in TV shows can also be created this way.

These first two photographs below show NASA using the Chroma key technique to create the filming of the Apollo Capsule's re-entry into earth's atmosphere and its parachute ride down to the ocean. Notice how the capsule drops below the ground level. This may have mimicked the splashdown into the water.

Below are a few more shots of NASA using the Chroma key special effect for the production of the Apollo Moon landing movies. Thanks to NASA and the American taxpayers' money, movies now have tremendous special effects. The effects, in some cases, are so good that nobody can tell the difference between a real adventure and a special effects adventure.

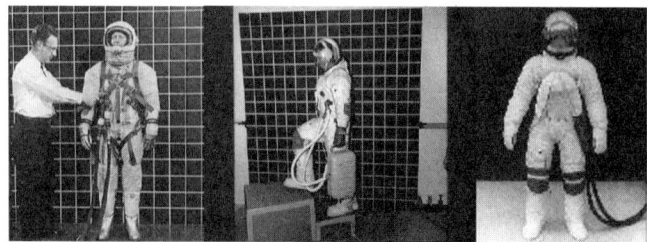

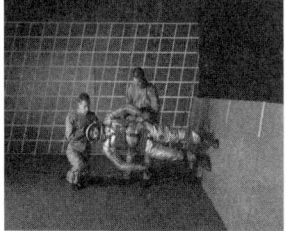

How NASA Simulated the Weightlessness of Space

The Horizontal Gravity Simulator

The Langley Lunar Crane was also used to simulate certain portions of the Moon Walks.

To simulate the low gravity of the Moon's surface, in a horizontal (sideways) position, the astronauts were suspended from the Langley Lunar Simulator Crane by several strategically placed elastic bungee cables. The angled panels, on which the man walked or ran, were offset just enough to simulate the sixth of Earth's gravity that is present on the Moon.

NASA's decision to use a Horizontal Gravity Simulator "To produce the Moon's 1/6 gravity" was a wise choice for several reasons. First, it was much more likely to go undetected by movie viewers who were looking for cables extending above the astronauts. Also, this Horizontal Gravity Simulator offered the Apollo Astronauts more flexibility than using a traditional Vertical (Upright) Gravity Simulator.

As amazing as it may seem, to create the illusion of the Astronauts being much lighter on the Moon's surface, NASA simply used a modified "Cliff Climber" gravity reduction technique. With this system, it was very easy for someone to push themselves off the side of a wall while being supported by cables from above. The astronauts appeared to be effortlessly moving around on the Moon's surface, and that was how they managed to perform the most realistic looking "Moon Walk" ever imaginable.

Again, the unusual design of this system explained why, for over three decades, none of the skeptics of NASA's Moon landings managed to find any cables that allowed the astronauts to bounce around on the Moon's surface. Everyone has been looking for bungee cords above the astronauts in the upright position. Below are a few of NASA's photos where the Apollo Astronauts were hooked up to the Horizontal Gravity Simulator. The purpose of these photos is to illustrate where the cables were strategically placed for future reference to the upcoming alleged Moon landing photos that also have these cables hooked up to the astronauts.

How did NASA manage to get the Moon's surface and background scenery around the astronauts? That's the easy part. NASA again used the video special effects technology known as Chroma Key (a.k.a. blue screen).

In the next picture, we witness NASA filmmakers using the video special effects technique in combination with the Horizontal Gravity Simulator. Here they have added a special Chroma Key grid behind the astronauts and a black curtain to represent space. This allowed the astronauts to be placed in front of any fake background, or scenery NASA wanted.

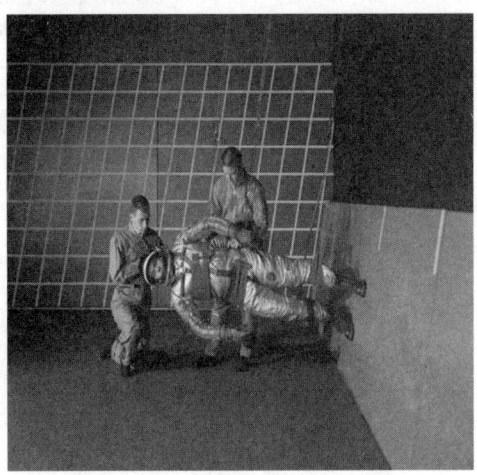

Using this combination system, the astronauts were seen in many imaginary situations, such as running and jumping on the moon. In addition, they were also seen, in some cases, lifting heavy objects effortlessly. With this system, there was no limit to what the astronauts could do. If NASA wanted the astronauts to run up or down a hill, they simply had them hooked up to one of the Horizontal Gravity Simulators with a curved Chroma Key grid. Below is a perfect example of a moon hill scene that matches the above angled Chroma Key panel. The combination of these two systems explains how NASA managed to produce the most realistic looking Moon landing movies ever imaginable.

This only costed NASA $56,611,332,000.00 (over 56 Billion Dollars). The Moon was not a complete waste; this special effect technique, improved by NASA, is now used so often that folks don't realize it. Complete segments in TV and Movie shows are created this way. Can you imagine how incredible the movies of the future are going to look after approving President Bush's new budget request to fake another manned mission to the Moon? Talk about exciting.

Inverting the Horizontal Gravity Simulator training photos reveal a whole different picture.

By turning the Horizontal Gravity Simulation training photos upright, they revealed many clues to NASA's Moon landing photography.

For example, the training photo revealed the astronauts running on the Moon's surface had the same characteristics of the astronauts running on the Horizontal Gravity Training Simulator. Notice how the astronaut's back leg was being pulled unusually close to the path of the extended front leg by the bungee strap. If the astronaut was indeed running on the Moon's surface and not just bouncing off the side wall of the gravity simulator, he would have fallen flat on his face.

For some combinations of astronaut scenes, it required each astronaut to be connected to his or her own independent overhead levitation system. This explained why there was a certain amount of distance between the astronauts, and why one was following the same path of the astronaut in front of him. They were running along the Chroma Key grid tracks with not much room to move from side to side. The above Horizontal Gravity Training portrait is a perfect example. It's only about 4 feet wide and has a curve at the end, to simulate the astronaut running down a hill.

As would be expected, the Apollo astronaut's moonwalk film footage contains an abundance of evidence that the vertical gravity simulator was being used to create the illusion of the Astronauts being on the moon surface with a 1/6 of earth's gravity environment. Especially during the scenes where the astronauts are running, jumping, and walking up and down hills. That's right, NASA's moon landing film footage confirms the use of the vertical gravity simulator.

Several obvious clues revealed the astronauts were actually hooked up to this Horizontal Gravity Simulator. Some of the hints were the abnormal positioning of the astronaut's legs and arms as mentioned before. This was a significant clue. Another obvious clue was the harnesses found attached to the astronauts while they were supposedly on the Moon, and signs of artist's air brush touch up work performed over the astronaut's clothing where the Vertical Gravity Bungee Strap had been attached.

Another obvious visual confirmation was the scenes where the astronauts were pretending to fall on the Moon's surface, and miraculously they bounce back up as if they experienced lighter gravity on the Moon. This is also a very important clue in proving the Moon landings were faked. While connected to the Vertical Gravity Simulator, what was involved in faking a fall was much different then actually falling on the Moon's surface. When the Apollo Astronauts were instructed to fake the falling on the Moon's surface, this was a big challenge. When someone falls on the ground accidentally, his or her feet slip out from underneath him or her as gravity takes over. However, when the astronauts were hooked up to the Horizontal Gravity Simulator, NASA's engineers overlooked one little problem. When attempting to falsely trip, or fall down from a stand still position, the astronauts actually become 6 times heavier.

Newton's Law of Gravity Proves Moon Landings Were Faked

We can now use Newton's Law of Gravity to prove the astronauts were not on the Moon during the filming of the Apollo Moon landing missions. When attempting a free fall from the Horizontal Gravity simulator, gravity actually works against astronauts, and they become heavier. Instead of being 5/6 lighter, the astronauts actually become six times heaver.

With the enormous amount of extra gravitational pull against the astronauts, rather than just falling over they were forced to first bend their knees and then push themselves off the simulator wall before they could fall over. If Newton's law of Gravity is correct, the astronauts' irregular motions would certainly have been seen in the Moon landing film footage, and it is. This is exactly what we see the astronauts doing in the Moon film footage.

In the motion picture, before falling, the astronauts repeatedly pushed themselves outward as opposed to simply slipping and falling. Once again, with a little investigation and some well-known basic scientific principles in hand, NASA's Moon landings have been proven to be fake.

Below are a few of NASA's photos that reveal the side mounted bungee straps attached to the astronauts while they were supposedly on the Moon's surface. What was NASA thinking?

NASA photo #: as12-48-7071   NASA ID # h_astronauts_si_02

Evidence Of Cutting and Pasting Photos Together To Hide Levitation Harnesses

Many of the Moon landing photos of the astronauts showed NASA's photographers were cutting and pasting photos together to hide any evidence of the levitation brackets being used. This discovery provided additional confirmation that the Gravity Simulators was where the actual filming took place for the Apollo Moon Landings.

Below is an example of the type of cutting and pasting that was done on NASA's Moon landing images to hide the Vertical Gravity Simulator's harnesses.

A Few More Movie Special Effects NASA used to Fake Moon Landings

The Moon Craters Were Man-made

NASA used several creative ways to create moon craters; one was to build crater simulators. Moon craters will be covered in more detail later in the book. For now, here are a few examples of how NASA's moon crater simulators matched the moon surface photos.

The sand they dug out of the hole can be seen on the moon surface in the above moon photo.

Miniature Stage Props

Many NASA shots showed miniature objects. Furthermore, these objects were used to construct moon landing scenes. For instance, in the next picture, there are two identical objects. Nevertheless, what would be the reason for the two different sized flags in the picture on the right? The most logical answer would be that the smaller one was used to distort the actual distance of the background scenery, which usually was less than thirty feet back from the camera.

A great deal of evidence suggests NASA used miniature stage props such as flags, lunar modules, and moon rovers. Subsequently, NASA pasted the background to the back of these photos.

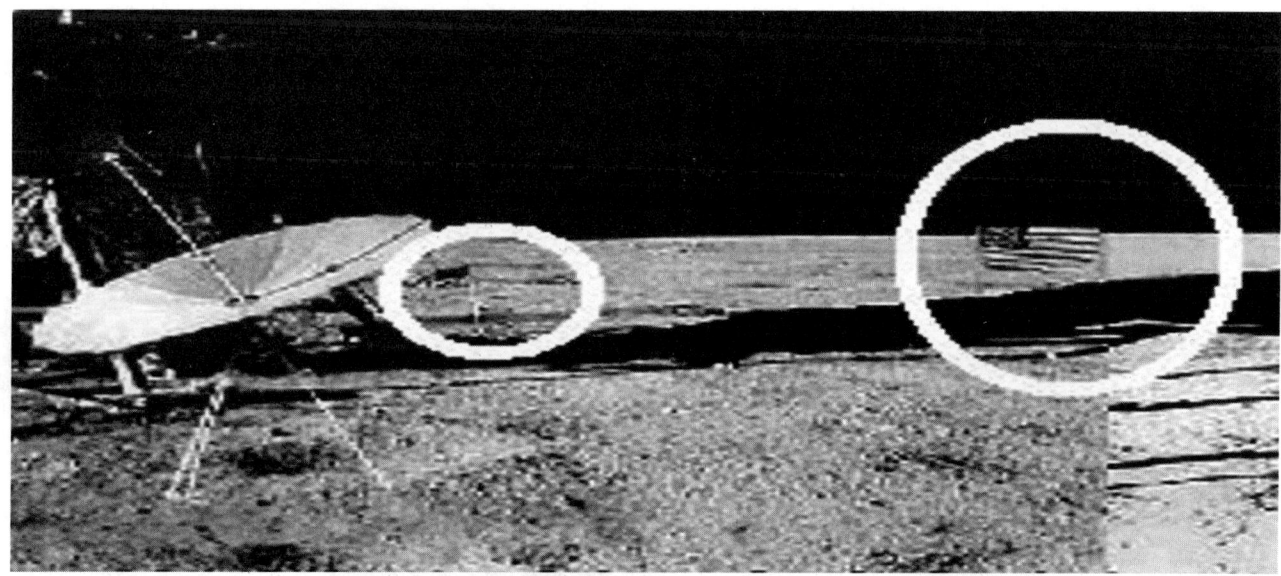

Extra miniature flag left on stage for this scene

Here is an example of how miniature stage props were used to create fake moon landing photography. To produce these moon landing photos, it appears NASA used a miniature LM spacecrafts in combination with a painted backdrop.

NASA

In the 1960's, this was a very common trick. Watch any Godzilla or King Kong movie, and you can see how easily movie producers distorted the actual distances and sizes. Applying the same tactics, NASA artists could easily have pasted earth's mountain scenes in the background, blacken the sky out, and there you have it. A photo taken on the moon.

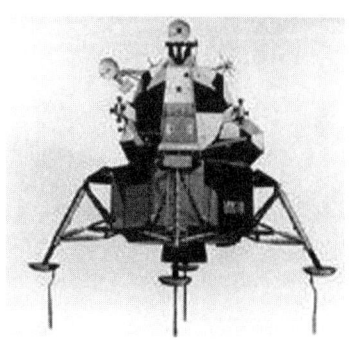

NASA Astronauts Used A Movie Script And Pre-made Maps Of Landing Sites:

If the astronauts weren't really actors, how did they seem to know exactly what to say and do? If the hoax theory was to stand up to scrutiny, this question needed to be answered.

An anonymous source claimed that NASA attached movie scripts and maps to the actors' arms with custom wristbands. As we searched for clues to confirm this rumor, we did discover several NASA photos that indeed showed these pre-made scripts and maps. These movie scripts were found in earth and moon pictures. See for yourself! Look below!

With these movie scripts, the astronauts (actors) did not have to know much about being an astronaut. The astronauts could simply show up to the training simulator movie stage. Next, they followed the filming instructions attached to their arms for that day. There are no questions about it. This is exactly how they did it. The maps even appear to match the terrain the astronauts were supposedly going to encounter later on the moon.

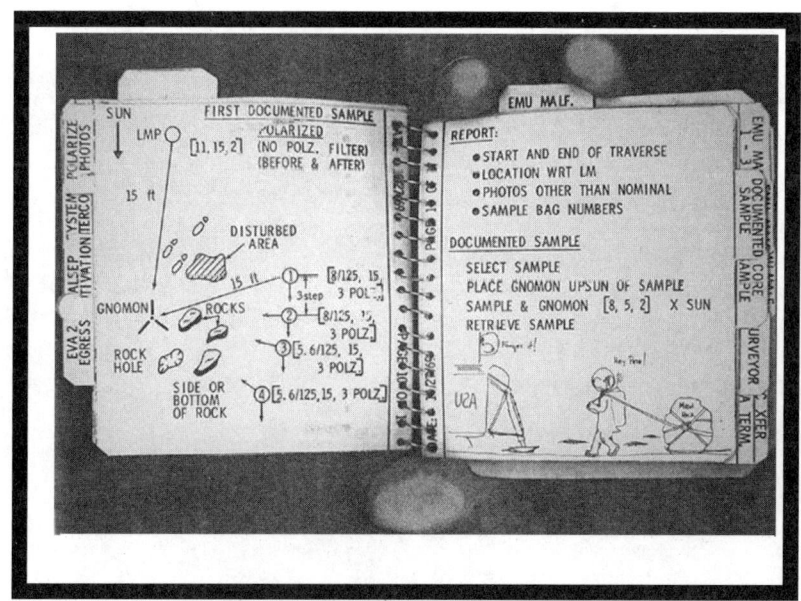

6-29

Chapter 6- Part 3

NASA Used Look-A-Like Dummies As Astronauts

To fake the moon landings, NASA used look-a-like astronaut dummies! Is that the most ridiculous thing you have every heard? Actually, this should not be hard to imagine. Look-a-like dummies date back to the ancient civilizations of Egypt and Greece. Historical documents confirm that dummies were sculpted during the Roman Empire for religious ceremonies. During the Apollo Moon landings era, before digital animation evolved, look-a-like dummies were used heavily by the movie industry. Further, they were used in dangerous scenes. Some of the scenes had **mannequins** (look-a-like dummies) hanging on to the back of fast moving cars. Other scenes had dummies strapped to horses for very fast and bouncy rides.

These movie stunt dummies were so realistic that people couldn't tell the difference. A perfect example would be the famous movie, 1959 Ben-Hur. It's often claimed that a stunt man was killed during the filming of the chariots being dragged by the horses. In the scene, a man suddenly turns to face the oncoming chariots. Next, he leaps out of the way, and thereafter is run over by a second chariot. Before the man is run over, the camera cuts away to a posed dummy, which gets flattened. Many people point to this as the fatal accident. People also claimed that during the taping of another scene a Roman soldier was slaughtered by the same chariots. When a person views the movie in slow motion, even after being trampled over, you can see the body's legs remain straight. Note particularly that the feet remain perpendicular to the legs. These were simply distant shots all done with stunt dummies, yet the media had reports that stated these stuntmen were killed.

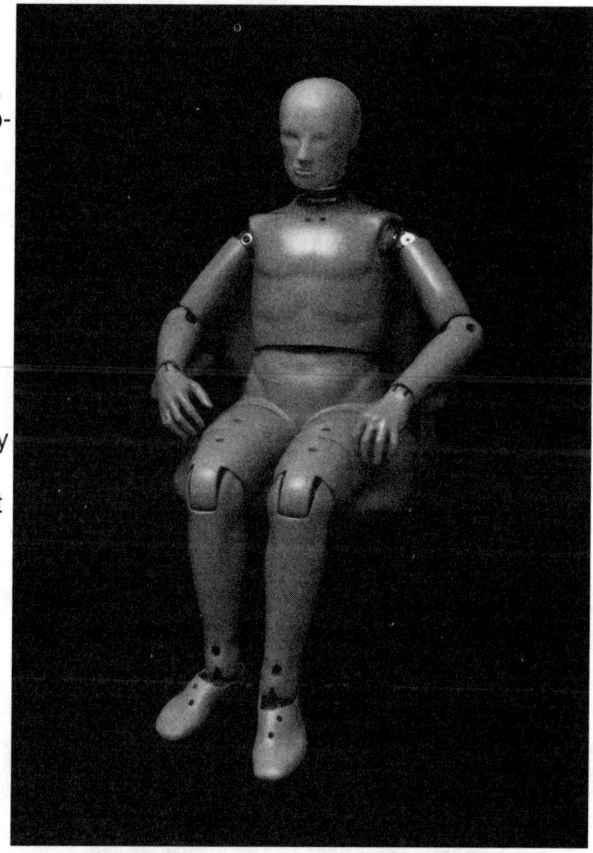

While analyzing the Apollo mission photos it was clear that NASA was presenting many photos that showed conditions simply impossible for astronauts to have preformed in space, or on Earth. For instance, there are many photos of the astronauts fastened to the Moon surface and the LM. There are bolts and wires running through their arms and feet. There are pictures with large tears in their spacesuits. There are even many photos showing birds, cats, and insects with the astronauts on the Moon surface.

Only after considering the Moon landings as having been faked here on earth, did everything start to make sense. In rare instances NASA's movie producers used look-a-like dummies (mannequins) to represent the astronauts. This may sound hard to believe but when analyzing the photography NASA claimed was taken on the Moon, it's the only thing that makes sense.

How NASA Simulated the Weightlessness of Space

The following are a few of the Apollo Moon Landing photos, which under close examination, reveal the use of look-a-like dummies. A majority of these photos were taken at the Kennedy Space Center and Langley Research Center's elaborate moon landing training simulators (movie studios).

The Astronaut Look-a-like Dummies Were Fastened To the LM Spaceship And the Moon Surface.

With a magnifying glass, one can clearly see this alleged photo of the astronaut on the moon is actually a look-a-like dummy. In order to maintain the astronauts position for the precision photo shot, harnessing clamps were inserted into the dummy's foot .

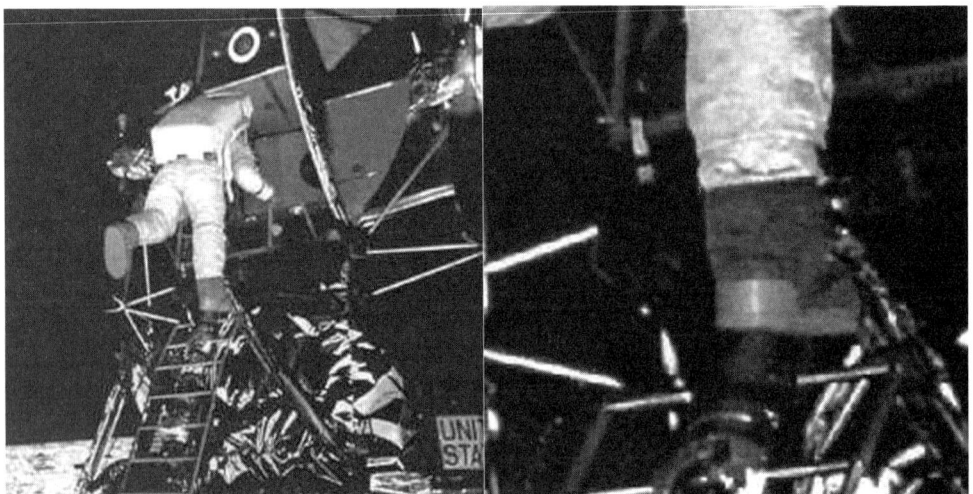

This is a look-a-like dummy with a support rod through its foot
NASA ID# as11-40-5866

With limited photographic technology available in the 1960's and 70's, correcting original film footage using dummies was the method of choice for reproducing an exact scene. It would have been impossible to keep an astronaut levitated in mid air to re-create, perfectly of an earlier position.

NASA still uses the same look-a-like astronaut dummies in many different ways. It is amazing that no one has ever put the connection together. When comparing the Moon landing photo below with the Earth museum picture beside it, it is easy to see how NASA substituted dummies for astronauts. In both photos the astronaut is strapped to the ladder, which would not have been necessary on the Moon.

Boot strapped at Bottom

NASA ID# AS11-40-5868

This new hoax theory needs to be put to the test. Could these photos simply be a camera somehow playing tricks on us? Once again, after inspection of thousands of Apollo Moon Landing photos, additional evidence was uncovered. For example, the look-a-like dummies were found supported by wires in space and on the moon surface. There were snapshots showing astronauts without life support. The image also revealed tears in space suits while "in space", and the astronaut's helmets were opened. Because of this, dummies were visible. The list of proof goes on and on. In the next few pages of NASA Moon landing photos, are highlighted some of the best.

Wires And Clamps Supporting the Dummy Astronauts

The pictures below are just a few of the many museum's displays where look-a-like dummies were being used as substitutes for astronauts. In almost every aspect of the moon landing missions, the dummies were used to simulate the astronauts. If NASA actually had look-a-like dummy technology back in the 1960's, to help produce a very realistic moon landing hoax, they definitely could have used this expertise in many different ways.

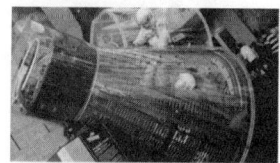

Astronaut's Helmet Found Open On t6he MOON Surface.

NASA must have kept this. They certainly would not want it to be known one of their astronaut's died on the moon. For such a stupid mistake as jumping out of the ship with his helmet wide open.

NASA ID# 20117422 Open Helmet

Wide Open Helmet On the Moon

More Open Helmets

NASA ID# 10076104

The Discovery Of the Use Of Look-a-like Astronaut Dummies Also Helped Confirm the Use Of Outdoor Movie Studios (Moon Training Simulators) To Stage Portions Of the Space Walks

To confirm this theory, a thorough search was conducted. Further, before releasing the moon landing photos, we looked for evidence that may have been missed by the NASA photography editors team. Indeed, one could imagine pictures of animals that managed to get recorded.

The outdoor movie studios theory was confirmed with an abundance of proof. There are so many NASA Moon landing photos showing birds, cats, dogs, and other animals. For this reason, an entire chapter of this book was dedicated to animals found in the alleged Moon landing photography. For example, take this space walk picture. A field cat has decided to sleep on one of the astronaut's arms. On the other arm, a bird is resting.

Bird and animal Resting On A Look-a-like Dummy's Arms

NASA ID# 10074015

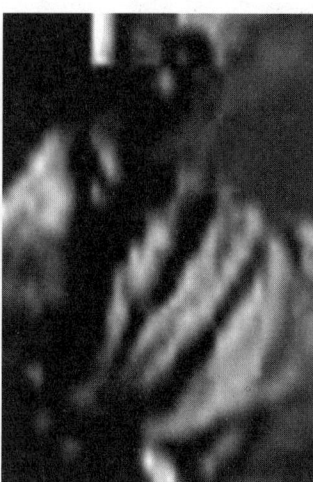

Cat Resting On Look-a-like Dummy's Head

NASA ID# 10073969

Snake Slithering Out Of Astronaut's Open Helmet In Outer Space

NASA ID# 10074584

Yes, as strange as it may seem, these pictures make sense because, for many years, filming took place in the same locations. Over time, an abundance of animals dwelled on the outdoor moon landing studios. In fact, a few creatures even resided there. This is why a great deal of filming was done inside NASA's manned space centers. In addition, inside the centers, was less interruption from animals and weather conditions.

Cats were apparently the biggest problem for the moon hoax movie crew. The artificial moon dust surface used outside created the Worlds Biggest Litter Box for cats. It was impossible to keep the cats away from the outdoor movie studios. Consequently, cats left evidence in many moon landing pictures. When the original moon landing photos were closely explored, many cats and other Earth creatures were seen in tricky positions. This explains why these animals were not airbrushed over to look like rocks. The animals were undetected by NASA editors. To illustrate a perfect example, refer to the famous "C rock". After being discovered, NASA's artists must have quickly touched up the rock and removed the letter "C". Evidence suggests that NASA constantly converted and edited over items, such as birds and snakes, to hide their appearance.

Siamese Cat on Moon Surface.
NASA ID# 20124287

This next illustration shows a cat on the movie studio outside the Langley Research Center. The animal is sleeping in an almost unrecognizable posture to the human eye. If you look closely, you can see the wild cat sleeping on the dummy astronaut on the left.

Cat Sleeping On Dummy Astronaut Stage Prop.

Cats Are Discovered Hiding Between Two Rocks And Moving Freely Around On the Moon Surface.

Cat

How NASA Simulated the Weightlessness of Space

Cat Walks With Astronaut. This snapshot came from an anonymous source; however, everything indicates it is a NASA original.

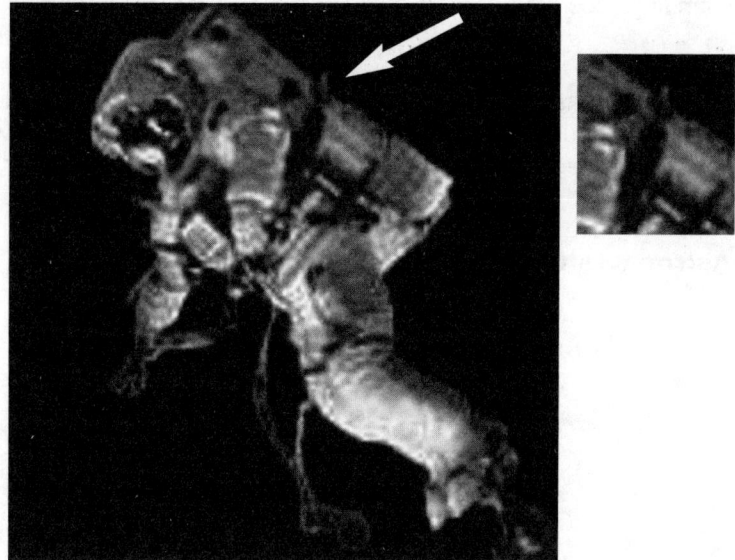

6-37

After searching NASA's records, evidence was uncovered proving that NASA had often used look-a-like dummy astronauts before and during the Apollo Moon landing projects. For example, during one Pentagon news conference in June of 1997, NASA admitted to using human look-a-like dummies as early as July 1947. This was long before the alleged moon landings occurred.

As far as the alleged manned missions to the moon, these next pictures confirm that dummies were used. NASA claimed the three look-a-like dummies seen attached to the spacecraft's couch were used for the Apollo Moon Landing projects. These mannequins are referred to as apollo dummies. Further, even up close, they are easily mistaken for astronauts.

Linking the Apollo Astronauts to Using Look Alike Dummies

NASA Apollo Astronaut Look-a-like Dummies

US Air Force Photo - Apollo Project 1965

One last thing that still needs to be done to close this piece of the Moon landing hoax puzzle, is to link the Apollo Astronauts to the look-a-like dummies. In order to establish a solid connection to this hoax theory, three very important things need to be answered.

First, evidence must prove that NASA put dummies into the Apollo Astronaut's spacesuits.

Secondly, NASA should testify that, to replicate space mission scenes, they used dummy astronauts.

Thirdly, it ought to be proven that the Apollo Astronauts had knowledge of these look-a-like dummies.

Astronauts watching Dummies NASA ID#i8-12a.jpg

After searching thousands of Apollo Moon Landing mission records, one special photo seemed to perfectly answer all three very important questions. The shot was taken during a press conference in the Manned Spacecraft Center and included several of the Apollo Astronauts. It is important to note that this photo was taken on April 23, 1966 before the astronauts alleged trips to the moon.

If A Picture Could Say A Thousand Words

If a picture could say a thousand words, then this next one is for the record. Look at their faces. Furthermore, there is no doubt everyone of these so-called astronauts knew look-a-like dummies were being used in their place. As an example, stare at the image to the left. What was NASA thinking? It is hard to tell for sure; however, it's believed that the look-a-like dummy seen in the background is Buzz Aldrin's.

Look-a-like Dummy In Spacesuit

NASA ID# 10074328

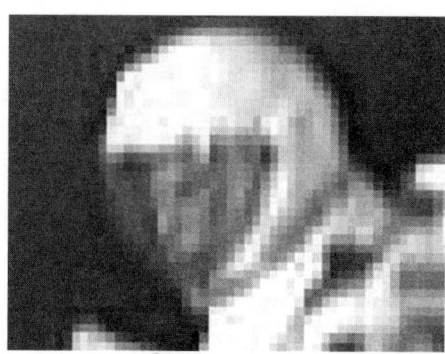

Close-up of
Look-A-Like Dummy Astronaut

6-39

CHAPTER 7

How NASA Faked the Mission Equipment Capabilities

NASA-Dryden photo

Lights, Cameras, Action

Chapter 7 - Part 1

How NASA faked the Lunar Lander Training Vehicle Tests

Chapter 7 - Part 1
How NASA faked the Lunar Lander Training Vehicle Tests

Remember how twelve of the original astronauts that tested the Apollo space equipment died in mysterious accidents? This chapter is going to attempt to provide a possible answer for those mysterious deaths.

To help solve this mystery, we looked at the roles these astronauts would have played in the Moon landing hoax. Their primary task was to test the Apollo moon landing equipment, so first, lets see if there were any problems with the equipment.

On the surface, everything appeared to have been normal and the test filming looks great except for normal trial and error problems. It seems strange that several of the Apollo testing astronauts seemed to contradict NASA's claims of perfection. For instance, astronaut Gus Grissom constantly referred to the Apollo equipment as a "lemon." He meant that the equipment was totally unreliable. Gus even left lemons on the Apollo gear as clues, so earthlings of the future could one day discover and reveal the truth.

According to Gus wife, Gus was scared for his life. Just three days before he died, in a freak Apollo training exercise accident along with two other Apollo Astronauts, he told her he was going to be killed over the Apollo project. Betty Grissom, has long said NASA covered up what really happened in the launch pad fire that killed her husband and fellow astronauts Ed White and Roger Chaffee during an Apollo 1 dress rehearsal on January 27, 1967. What she calls the "murder" of her husband.

Because the Apollo Astronauts' reports totally contradicted each other about the Spacecraft's capabilities, I felt it would be a good idea to study the major components of the Lunar Landing Training Vehicle (LLTV). An in-depth examination applying basic principles of modern aeronautics should determine which astronauts were telling the truth, and whether or not NASA greatly exaggerated the spacecraft's capabilities.

After a thorough investigation of the Lunar Lander mechanical design, aside from the fact that it was a mechanical disaster, the results indicated there was something very strange about the spacecraft. Our research and supporting evidence indicated that, in almost every way imaginable, NASA totally fabricated many detail of the entire spaceship! Our findings suggested that the ship was clearly incapable of flying, as the fashion NASA claimed it could. As we can see in the visual observation below, as well as in many other films, the Lunar Lander appears to be flying effortlessly under its own power with the guidance system functioning perfectly.

Again, there is a serious problem with the ship flying, our research determine the Lunar Module Spacecraft could not fly under it own power, as seen in the above NASA training photos.

How can we have totally opposite points of view? NASA and I are basing both of our theories on basic aeronautic and physic principles, which are relatively constant? Something does not add up here. There seems to be a similar situation with the astronauts contradicting each other's opinion of how the Lunar Spacecraft operated. What is going on here? Why would some Apollo Astronauts claim that the equipment worked perfectly, yet others implied the odds of success were zero and any attempt would have failed?

Furthermore, some NASA engineers who worked on the Apollo project, have admitted in books they have written that the Apollo spacecraft was plagued with hundreds of serious defects. If NASA engineers admitted there were hundreds of very serious problems, why didn't all the astronauts admit to the serious troubles? Were some astronauts kept quiet out of fear? Look what happened to the Apollo Astronauts that talked? They later died in mysterious deaths, while the astronauts that claimed the spacecraft worked perfectly lived long lives. Sue Lynn Chung, a young Internet Whiz kid from China, came up with the answer that solved the Apollo capability controversy, and it's a good one. Sue Lynn suggested that NASA had two separate groups of astronauts. One group tested the actual Apollo equipment, and another group of astronauts (actors) filmed fake flight tests at a different location. After putting Sue Lynn's theory to the test against NASA's own evidence, it was discovered that a great deal of evidence, including NASA's photos and mission records, confirmed Sue Lynn's theory.

If Sue's theory was correct, the first group of astronauts testing the actual equipment would have never been told by NASA that they had no intentions of sending them, or the equipment they were testing, to the moon. Because these astronauts were working directly with the rocket design companies, NASA could not take the chance of them accidentally telling someone that the moon landings were being faked. This makes sense. If the first group of astronauts had known about NASA's plans to fake the moon landings, they would have kept quiet about all the problems they were experiencing with the equipment, yet they did not. Remember how Gus claimed he felt the equipment functioned so badly it was going to kill him?

Great deals of evidence, including NASA's illustrations and mission records, support the theory of two separate groups of astronauts. The best proof is the actual Lunar Lander Training Flights themselves. Let's re-examine the four Apollo Lunar Lander flight pictures taken back in the 1960's and 1970's. When comparing these flight test pictures to the photos of the actual firing of the rocket engines, which were supposedly mounted inside the Lunar Lander Spacecraft, something was seriously wrong.

Rocket Engine

LM Spacecraft Rocket engine

Notice how the two pictures on the left have the amount of engine thrust similar to a smoke bomb, and the other two Lunar Landers appear to have no engine thrust at all? Obviously, the Lunar Landers were not flying under their own power, but how did NASA accomplish this feat? As seen in the images above, these test flights were filmed outdoors in a wide open area with absolutely no crane seen to levitate the spacecrafts.

If the Lunar Lander Spacecraft could not fly this in this fashion based on all the evidence the found. Then a second group of astronauts must have been separated from the real astronauts and did the fake filming at a different location. The search started where the actors were working. How could NASA manage to make the spacecraft appear to fly without being detected in the film footage? If Lynn's theory was to be accepted by the scientific community, these questions needed to be answered. The good news is Sue Lynn's theory has been verified, and it was, by far, one of the most incredible discoveries yet.

All data suggests a second group of astronauts (actors) were most definitely filming fake spacecraft flights at the Edward Air Force Base in California and at the Ellington Air Force Base in Texas. During the 1960's and 70's, these air force bases combined were believed to have more movie equipment than Hollywood.

Locations Where NASA Faked the Spacecraft's Ability To Fly Under Its Own Power

To simulate the Apollo Spacecraft's ability to fly, the following series of pictures reveal one of the methods used by NASA. Although it may seem a little strange, it is a very simple concept that worked perfectly. The spacecraft was levitated in the air by a specially designed helicopter. Hence, the Apollo Spacecraft appeared to have the ability to fly. The following portrait shows how this heavy duty NASA helicopter was customized with special harness brackets to hoist the spacecraft and maneuver it around for the moon hoax filming.

This first picture is of a helicopter specially designed by NASA hovering above the Apollo Lunar Lander Spacecraft. It seems as if this jet was just about to pick the spacecraft up and fly it around. At first, we thought this was just a coincidence. When this picture was taken, a helicopter just happened to be passing by. The more we examined this picture, the stranger it looked. The first thing we noticed was the helicopter was like nothing any of us have ever seen before. It had two extra large propellers. That helicopter was clearly designed to carry one very heavy load and still maintain stability.

This next NASA picture helps solve the mystery.

With the helicopter hovering above it, the spacecraft was beginning to look a lot more like a levitation bracket designed especially for the spacecraft to be picked up and maneuvered around by a helicopter. Could this be a picture of the helicopter as it just arrived to the filming location? At this location, it would fly the astronauts around. Thus, everyone would be fooled and they would believe the ship could fly on its own.

Maybe this is a picture of the helicopter landing the spacecraft. Perhaps NASA's team of artists removed the lifting cables. This was a question that needed to be answered and prompted a further investigation of all NASA's other spacecraft testing photos. Could the Lunar Lander Spacecraft, seen in all the flight tests, be nothing more than a levitation device used like a stage prop? If this helicopter was used to fake the Lunar Lander Spacecraft's ability to fly, then like all the other moon landing hoax questions put to the test, other NASA facts would confirm this theory.

Confirmation That A Helicopter Was Used To Fake the Spacecraft's Flying Capabilities

In an attempt to confirm this new helicopter Moon landing hoax theory, the picture to the right is composed of two different NASA Apollo training photos used for experimentation purposes. When studying the two vehicle design features, it is clear that they were both designed specifically to facilitate the act of flight. However, at this point, this is still only a theory. The next step is to see if any evidence supporting this hypothesis can be found in NASA's official Apollo records.

Sue Lynn, and the Internet Whiz kids of Japan, discovered evidence that NASA was faking the spacecraft flights with the helicopter at Dryden Flight Research Center. It was amazing how quickly these Whiz Kids from Japan found the evidence. As they searched feverlessly for some type of clue, they must not have slept for days. When Sue Lynn and the other Japanese Whiz Kids were asked why they were working so hard to help solve the moon landing hoax they said, "hopefully when this book is published it will convince President Bush and NASA to admit to the Apollo moon landing hoax". Until then, they felt everyone's scientific advancements were being held back. Sue Lynn said, "A lie like this should not live forever".

We could go into a great deal of technical design specification information that proved the LLTV was incapable of flying under its own power. Nevertheless, that really isn't necessary here. In this case, a few pictures speak a thousand words. The photos below clearly show the helicopter crew hooking up the hoisting cables to the Apollo Lunar Lander Spacecraft. Before being flown around, the astronaut (actor) took a photo. Later, NASA photographers would edit out the cable harnesses and airbrush any signs of foul play. In the picture on the right, notice how the harness cables were strategically attached to the Lunar Lander Spacecraft.

Hoisting Mechanisms for LM spacecraft

NASA ID # ap11-s69-39265

NASA ID # ap11-s69-36910

The above figures on the left exemplify the helicopter pilot inspecting one of the levitation brackets on the helicopter, which were attached to the levitation brackets on the Lunar Module Spacecraft just above the ladder. In the picture below, on the right, as the spacecraft was prepared for its helicopter flight, the levitation brackets and straps are clearly seen attached to the spacecraft. It's hard to imagine that these levitation cables, attached to the spacecraft, could be some type of secret flight technology developed by NASA, which assisted the LLTV spacecraft to fly on its own.

Just because everything pointed to the likelihood that NASA was using the helicopter, none of these portraits actually show the fraudulent act taking place. Additionally, since most of the photography related to the Apollo moon landings were locked up and protected by 24 hour armed guards, we have very little evidence to work with.

What makes this particular task even more difficult is the fact that something so obvious, like a helicopter flying around the astronaut in the spacecraft, was never going to get by NASA's watchdogs. Because of this, the photos were never released to the public. All we have to work with is a few graphics that have already been touched up by the NASA airbrush artists.

Well, maybe we will get lucky and find something showing NASA in the act of faking the spacecraft flying capabilities. Let's start with this NASA photo. Further, maybe it can provide us with a clue that NASA's spacecraft was incapable of flying under its own power. It was supposedly taken from a scene where one of the Apollo Astronaut's powered up the spacecraft engine, and took off flying around flawlessly.

Was the spacecraft truly flying under its own power, or were the filmmakers simply positioning the cameras at special angles to make it look as if the ships were flying under their own power? The filmmakers could have. To remove the cable harnesses and helicopter shadows, it would have only required a little bit of an artist's touch-up.

Dryden Flight Research Center ECN-453 Photographed 1964
Joe Walker pilots the Lunar Landing Research Vehicle (LLRV). NASA photo

To prove whether this astronaut was actually flying the spacecraft, the easiest way would be for NASA to declassify all of the evidence stored at the Johnson Space Center, which is protected by armed guard. However since that realistically is unlikely to ever happen, we will have to rely on the limited evidence currently available to the general public. To determine if NASA faked this spacecraft flight, two important questions had to be answered. First, in this portrait, was there any evidence indicating this ship was not flying under its own power? Secondly, can such evidence be found in all the other similar spacecraft flight photographs? The answer to both of these questions is YES! In this snapshot, there was a major clue that proved this spacecraft was not taking off under its own power. Furthermore, this evidence of fraud was found in all of the Apollo training flight photos available.

Although the clues in this picture were not so obvious, it was a very significant fact indeed. If you look to the lower left hand corner of the above picture, you can see a small portion of a truck parked directly beside the LLTV Spacecraft. If that astronaut even attempted to really start that rocket engine with that truck parked next to it, the rocket engine would have exploded.

This same truck was seen beside every other Apollo training flight we have found. It's even parked next to the LLTV Spacecraft with the helicopter levitation cables attached to it. Once again, indicating the ship could not have used its powerful rocket engine to take off without destroying the utility truck and possibly killing the astronauts on board. Since this truck can be seen beside the LLTV Spacecraft, in almost every test flight, it must have served a valuable function. Why would a truck be so close to a rocket ship while the ship was powered up and ready for take off? Did it serve as a barrier wall between the film crew and the spacecraft? That would explain how NASA managed to get right up to the spacecraft and was able to take so many spectacular looking close up pictures. Certainly, a more realistic explanation would be the utility truck was being used to help create the illusion of the spacecraft flying during the "Lift Off" portion of the flight.

NASA ID# as11-s69-36910
Custom Steam Truck behind LLTV

Based on the pressure gauges visible on the outside of the utility truck, one would have to assume it was some kind of high pressure steam generating vehicle. What would NASA be using a high-powered steam generator for? Surely, during lift off, if NASA wanted to fake the spacecraft's engine exhaust, all they would have to do is start the 10,000 lb thrust engines and there would be a huge exhaust plume. This can be seen in the engine testing illustration below on the right.

Steam Truck
Steam Truck

LLTV Rocket Engine Test

Wait a minute! Lets look again at the size of that engine exhaust seen during the pre-testing. It is pretty scary looking. Could it be that the Apollo Astronauts could not have turned on the actual engine after it was mounted to the Lunar Lander Spacecraft?

Could any astronaut actually ride on such a powerful engine directly above the exhaust plum without being shielded? All joking aside, you get the point. As soon as the astronauts attempted to start the spacecraft engine, they would have been turned to charcoal, or instantly vaporized. Can you imagine anyone even trying to stand 30 feet from that rocket engine when it was turned on? Anywhere between three to ten thousand pounds of engine thrust, with a temperature in excess of 1,800 degrees, would have consumed the astronauts. The shear amount of engine thrust levels was additional proof that NASA's claims were scientifically unsubstantiated, and were made up for their Moon landing hoax. Doubtless, the astronauts could not have flown the Lunar Lander under its own power, or they would have been burnt to a crisp.

Therefore, the high powered steam generating truck seen next to the Lunar Lander makes perfect sense. The steam truck was used to create the exhaust smoke for the Lunar Lander engines lift off and the flight film footage. The next set of NASA's Apollo training pictures, taken back in the 1960's, are a perfect example of how the steam generating truck was used to create the illusion of the Lunar Lander having the ability to fly. These portraits show the steam generating truck's exhaust hose placed under the Lunar Lander Spacecraft as it was spraying out steam.

NASA PHOTO ID# a11.lltv3

Notice how the engine was directly above the steam hose. This was done to make it look as if the spacecraft engine were running. When referencing this theory against the live NASA film footage of the LLTV Spacecraft in flight, it was easy to see in the famous John Glen training flight, the position of the rockets, in many cases, did not match the direction in which the LLTV spacecraft was traveling. Yes, NASA's live film footage also confirmed that the spacecraft was not flying under its own power, it was nothing more than movie special effects at best.

In an attempt to make it look as if the astronauts were flying the spacecraft effortlessly through the air, the astronauts simply maneuvered the engine's direction hydraulically to follow the direction the helicopter was taking them.

How NASA Faked the Mission Equipment Capabilities

Other Methods NASA Used To Assist In Making the Apollo Spacecraft Appear Capable Of Flying Efficiently Under Its Own Power.

Chroma Key- Green (and Blue) Screen. Here we saw NASA utilizing the Green Screen special effect again. The picture on the left was the Apollo Spacecraft being maneuvered over a massive Chroma Key- Green screen. Undoubtedly, the picture was the end result of a pretty good editing job. If it were not for the fact that the Sun was reflecting in the opposite direction on the spacecraft as if it were on the ground, it would be very unlikely that anyone could tell the second picture was a fake.

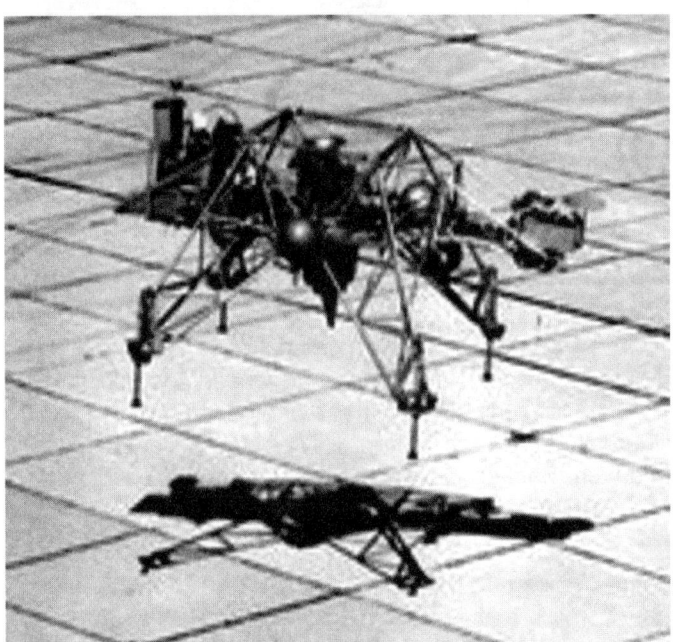
http://www.dfrc.nasa.gov/Newsroom/X-Press/1999/Oct29/images/flight.jpg

Lunar Landing Research Vehicle flies above the skies of Dryden NASA ID# LLRV20.jpg

NASA Also Used the Langley Crane To Stage the Spacecraft Flight For Their Moon Landing Hoax

As mentioned before, the Langley Crane served multiple purposes. It not only was used to fake the space walk, it was also used to simulate the descent and take off of the LM Apollo Spacecraft. Below are some snapshots illustrating how it achieved the effect.

2000-001902.jpg

EL-1996-00200.jpe

The concept of the Langley Space Flight Crane Structure being used to fake the Moon landing hoax, is far from an original idea. This type of crane structure has been used to simulate rocket space flights before, and NASA simply copied the idea from someone else.

The United States learn how to fake their Moon landings from the French

The next sets of pictures started showing up around the world back in 1955 almost 7 years before NASA started their Moon Landing project. At the time many people believed these were space alien UFO's or some new type of spacecraft the French air force hard developed.

UFO's Spotted Over France in the summer 1955

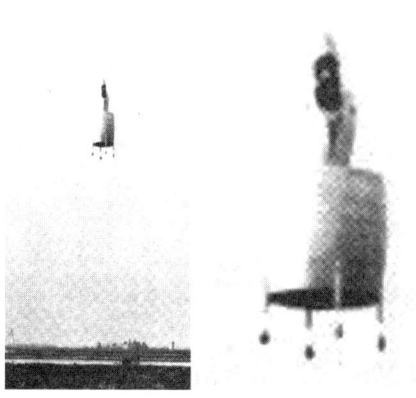 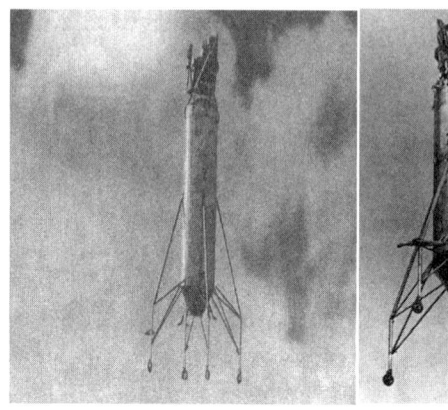

Of course these were not UFOs and France had not yet perfected this type of advanced space flight. Some one working forFrance's Aero Space engine company "SNECMA" distributed these photosas a joke. "SNECMA" had just started a test program with a spacecraft design called the "Atar Volant" ("flying Atar"), These spacecrafts could not fly under their own power and at the time were hooked up to a crane support structure. Although once the person painted over the harness cable, no one could tell the difference, as illustrated in the original picture to the right.

Once NASA realized in 1963 that a real trip to the moon was out of the question, they knew if they faked the mission they would have no problem with the actual spacecraft flight. Confident of how easy it would be to fool everyone with this type of space simulator, NASA made the decision to go ahead with the moon hoax. They began construction at a massive fight simulator at the Langley Research Center in Langley, Virginia, which was much more advanced than the French version.

 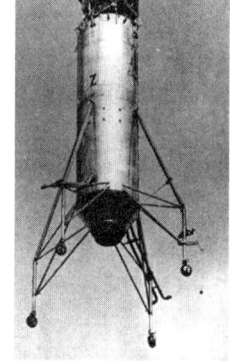

Original Picture taken   Cables covered

NASA's decision to use the French system of simulation space flight proved invaluable, until now no one has ever been able to figure out how the space flights were faked. Below is an actual picture taken of the crane structure NASA constructed based on the original French model and used to fake the moon landings. More details on how the Lunar Lander Simulator was used will be discussed later in this book.

Maybe the reason why France is angry at the America is because; they realize how NASA used their technology to fool the rest of the world that they went to the Moon and they can't say anything about it.

Here is Some More Advanced Proof That the Lunar Spacecraft's Exhaust Was Faked

Photography of the Lunar Module spacecraft's ascent (departure) from the Moon showed an exhaust plume from the engine. In outer space, the exhaust plume should not be visible. Typically, rocket engines used a fuel-rich propellant mixture, which resulted in un-oxidized fuel in the exhaust gas.

This fuel reacts with oxygen in the atmosphere to produce flame and smoke in the exhaust stream. In the vacuum of space, this secondary reaction is nonexistent. According to NASA's design specifications, the Lunar Module used a propellant mixture of Aerosine 50 and nitrogen tetroxide, which, in the absence of atmospheric oxygen to react with, produces exhaust gases that are nearly invisible. Moon landing films show an exhaust plume coming from the Lunar Module Spacecraft's ascent engine. This dramatization of the Lunar Module launch on the moon can not be considered scientifically accurate.

In addition, the sound of the Lunar Module descent engine should be heard in the Apollo audio, but there was no such sound. Also, how could NASA have installed a camera in the center part of a rocket booster engine and taken pictures without it being burned to a crisp? The answer is simple; it wasn't possible. If the engines were placed as shown, the film footage could not have been taken. You will see, later in the book, how NASA actually produced their "evidence" of the moon landings here on earth in an elaborate moon landing simulator.

The shot on the right was also taken at the Edwards Air Base where the helicopter was believed to be flying around in the Lunar Lander Spacecraft. This NASA image shows a Lunar Lander being levitated by a crane, which was surrounded by a group of movie producers. As they filmed, the actor simulated a test flight of the spacecraft.

Although this picture illustrates how easy it was for NASA to produce a realistic Lunar Lander test that could fool anyone, this picture was taken 29 years later for the movie: "From the Earth to the Moon". With this picture, we can confirm that the Lunar Lander was designed as a stage prop levitation device. The proof was right in front of us.

NASA-Dryden photo

This picture also helped depict how the cables would have been attached to the helicopter in flight. The helicopter theory was easy to confirm with the NASA film footage where astronaut, John Glenn, was supposedly flying one of these LLTV Spacecrafts around and crashed. It was so obvious the helicopter simply released two cables from one side of the spacecraft and dangled (held up) the spacecraft on the other two cables. Once John was ejected safely, the helicopter crew then released the other two cables and the ship crashed to the ground. Without a doubt, NASA's methods of trickery were not that difficult to detect. This is the original LLRV Apollo Spacecraft, in this picture, and the vehicle was located at Dryden in Building 4801 where the old welding shop resided. It could not fly then, and three decades later, it still can not fly as demonstrated in this picture.

Please note nobody is saying an aircraft, similar to this one, is incapable of flying. Nonetheless, NASA's spacecraft was simply a stage prop. Its engine was positioned incorrectly, and was supposed to be stationary with 90-degree vent pipes. To maintain any type of stability, the spacecraft required a minimum of two exhaust vents. This does not. Besides, there was clearly inadequate protection for the flight pilot. He would have been cooked. The only machine to fly like this, at the time, was the original British "Bedstead", and NASA did not even take the time to copy it's design specifications.

Chapter 7 - Part 2

How NASA Faked the Astronaut's LIFT-OFF Into Outer Space

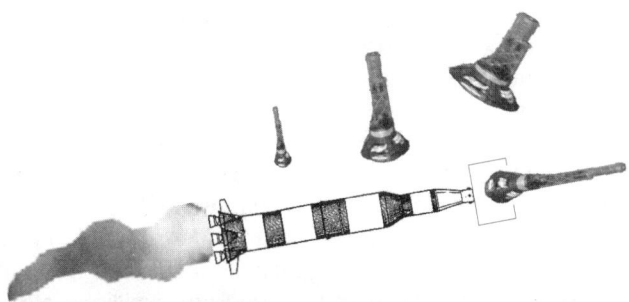

If the Astronauts did not go to the Moon, then what did NASA do with them?

If the astronauts could not have traveled more than 350 miles above earth's surface without being burnt to a crisp, then where did the astronauts stay during their mission time periods?

From just a glance, evidence suggests that the astronauts must have remained in outer space during their mission because there was certainly overwhelming evidence that confirms that the astronauts were launched into outer space. Most notably, NASA proudly proclaimed the event and broadcasted it live on television to hundreds of millions of people around the world.

This is, by far, one of NASA's strongest pieces of evidence that the astronauts went to the moon, and they have been milking it for all it's worth. For over thirty-three years now, NASA has been insisting the moon landings were real and referred to the fact that the Apollo launches were broadcasted live on television to hundreds of millions of people around the world. There were also hundreds of mission control staff and thousands of by-standers camping outside the Kennedy Space Center who witnessed, with their own eyes, the astronauts being blasted off into outer space.

Apollo 11 Liftoff Spectators
NASA ID# GPN-2000-001852

I don't know how you feel, but to me, based on this one piece of solid proof, it seemed as if NASA was daring everyone to try and dispute their claims. As if they believed no one on earth could ever find a hole in this one piece of evidence. Oh well, as long as NASA is making the offer, I would love to accept the challenge. Besides, if these launches were faked, isn't it about time the truth is revealed. If your wondering how I am going to find something different than the hundred of millions of people that have already watched the Apollo moon landing launched, remember, I am one of the few people that have been trained in almost every one of the basic principles that would be involved in a moon landing mission. The only reason NASA's moon landing hoax has gone undetected for so long is because most people have not been trained in all the fields related to a moon landing. That's not a bad thing.

However, this does explain why NASA feels there is no need to admit to their moon landing hoax. NASA realizes the only way anyone is going to figure out what cards are missing from their Apollo moon landing project, would be to go through the entire deck. Even if someone could figure out which cards were missing, they would still have to explain to others how the game was being played, incorrectly. The odds of this happening are almost zero. Why should NASA admit to faking the moon landings? Again, there are very few people still living that have the ability to go through NASA's entire deck of cards for the Apollo moon landings, and recognize just what cards are missing.

Now that we have a good understanding of NASA's motives behind the moon landing hoax, it's really not going to be that difficult to prove that the Apollo astronauts rocket launches were faked. NASA was sloppy with their evidence and left many clues to the hoax. One of the biggest clues that suggested the Apollo landings were faked is the fact that NASA deliberately broadcasted the event to the entire world. They did this during the heart of the cold war era. Considering the tensions between the Soviet Union and the rest of the world at the time, it's inconceivable that NASA would have immediately revealed to the rest of the world their new exploratory testing of the American's most top secret technology. Think about it. Broadcasting the Apollo moon landing missions, live around the world, would have been like giving the Japanese military leaders, in 1944, a tour of the "Secret Cities" manufacturing facilities where the first nuclear bombs were being developed. It's more likely the real reason NASA broadcasted the moon missions was to take advantage of the fact that most people were incapable of distinguishing the difference between a legitimate Apollo Astronaut rocket launch and a fake one. NASA simply realized early on that if they broadcasted the Apollo Moon landings, it would build a massive public false sense of belief in their moon hoax.

Even more suspicious than the broadcasting of the exploratory testing of top secret military technology is the government censoring the evidence. This contradicts their alleged manned missions to the moon. If you find it hard to believe the government censored this type of evidence, then why don't you see the videos the Whizkids discovered on TV? In addition, why don't you see the evidence contained in this book broadcasted on TV, or in the newsprint media?

When you really think about it, there were countless ways NASA manipulated the actual launching of the Apollo Astronauts into outer space. The challenge is trying to figure out what were the different methods possibly used by NASA's artists to trick everyone. Can any one of these be confirmed?

My first thought was that NASA had somehow constricted the viewpoint of eyewitnesses since they were restricting independent media coverage of the Apollo launches.

This next set of images makes it look like NASA removed the astronauts from the Saturn Rocket behind the launch pad curtain. It's obvious the astronaut's command module they rode in was removed from the rocket because the legs were extending outward at the bottom of the capsule.

What needs to be determined here is, was NASA simply removing the astronaut's from the rocket while they remained inside the capsule? Once the astronauts escaped, were they replaced with dummy astronauts?

As good as this theory sounds, with some further investigation, it was easily disproved. These were simply pictures of the command modules being installed prior to the astronauts being loaded on board on to the Saturn 5 Rocket. Furthermore, after reviewing several independent film sources of the Apollo 11 launch events and eyewitness reports, it was confirmed that NASA did indeed blast the Apollo 11 Astronauts on their way to the moon just as they claimed.

Confirming the Astronauts Remained On The Spacecraft During the Rocket Launching To the Moon Brings Up Another Question. Where did the astronauts stay during their alleged trip to the moon?

If the astronauts were launched into outer space, then they must have remained in earth's orbit until the end of their mission. Otherwise, any early re-entry attempt to earth would have been detected by independent radar sources, such as the Soviet Union's. Wait a minute! As plausible as that theory may seem, there was one major problem that prevented the astronauts from remaining in earth's orbit. According to NASA's original Apollo command module design specifications, there would not have been anywhere near enough life support capacity to sustain three astronauts in outer space for 8 days.

Because of this, if the astronauts could not have gone to the moon, or stayed in earth's orbit for the required time period, then through the process of elimination, the astronauts must have somehow remained on earth during their entire moon landing mission time frame. All I needed to do now was determine how the astronauts could have remained on earth. If this seems a little confusing, it's because that is exactly how NASA wants you to feel about the moon landings. If people were confused about the moon landings, they would take things for granted and often overlook the obvious.

There was a very simple answer to this piece of NASA's moon puzzle. Since there were almost 600 million eyewitnesses that confirmed the astronauts were on the rocket as it blasted off towards outer space, then the astronauts must have been on the rocket. Could 600 million people all have made a mistake? Besides, a double check of the lift off film footage confirmed the astronauts were on board.

The fact that the astronauts were confirmed as being on the rocket, at the time of lift off, made the task of explaining what happened to them much easier. Knowing that, at the time, NASA did not have the technical ability to support life in outer space, as you will learn later in the book, one has to conclude that the astronauts parachuted off the rocket as it raced towards outer space at over 6,000 miles per hour.

As crazy as that idea may sound, it's my theory, and I am sticking to it. Yes, the idea does sound far-fetched and very dangerous. Although, if NASA could have somehow removed the astronauts during the middle of the flight, before they reached outer space, in order to stage the return splashdown, NASA could have easily dropped the astronauts in the ocean from a cargo plane, or helicopter at the end of the mission.

If a sound method of escape for the Apollo Astronauts could be conceived and then proven to have been the system used by NASA, it would answer many strange and unexplainable scientific mysteries related to the alleged Apollo moon landing missions. Once and for all, everyone could put an end to the ridiculous nonsense of trying to come up with an answer to how NASA's Moon landing claims make sense. No more trying to explain why the flags were seen blowing in the wind on the moon, no more trying to explain why the Apollo Astronauts did not notice any stars in outer space. Why many earth creatures are found in the alleged moon landing photography. The list goes on and on! There were so many things that made no scientific sense related to the NASA moon landing claims, it is almost un-imaginable. If there were one thing that looks more suspicious than OJ Simpson's killing spree, it would be NASA's claims of landing the Apollo Astronauts on the moon.

After reviewing the Apollo 11 launch sequence and the design specifications of NASA's Saturn 5 Rocket, it was clear that there was a great deal of solid evidence proving the spacecraft was modified with a very sophisticated flight escape system. These modifications enabled NASA to safely remove the three astronauts from the booster rocket while it was racing towards outer space at over 6,000 miles per hour.

Understanding Basic Rocket Principles

In order to understand the method of escape NASA used to remove the astronauts from the rocket as it raced towards outer space at over 6,000 miles per hour, it is helpful to understand a little about how the Apollo Rocket was supposed to have functioned.

Heat Shield
Protects Astronauts inside of command module

When looking at the top portion of every rocket, you'll notice they are pointed. This section of the rocket is referred to as the "Nose Cone". Its smooth cone shape is designed to help the Apollo Rocket penetrate earth's ozone layer to reach outer space. The Nose Cone serves as a heat shield unit and was placed directly over the command module that the astronauts were riding in.

Command Module
Astronauts location during launch

The Nose Cone heat shield was supposed to function somewhat like placing a motorcycle helmet over your head. Passing through earth's ozone layer, from earth to outer space, would be like running from your garage to your house during a severe rainstorm with hail the size of gulf balls.

3rd Stage of Rocket
(service Module)

Imagine your head being the command module the astronauts were traveling in. During the hailstorm, with the helmet acting as your shield, the astronauts would have no problem going from your garage to your house. However, without the helmet to protect your head, the hail would certainly cause some serious damage.

The command module's only other heat shield was installed on the bottom for re-entry protection into earth's atmosphere. So, without the nose cone, while trying to leave earth's atmosphere, the command module would have been seriously damaged, and the astronauts would have been vaporized. It would have resembled the command modules seen in the NASA photo on the right.

Once the Apollo Rocket penetrated earth's ozone layer and safely reached outer space, the nose cone heat shield was no longer needed. The rocket design specifications then called for the nose cone to be jettisoned (removed) away from the spacecraft just as you could remove your motorcycle helmet once you have entered your home. The hailstorm outside that represented earth's ozone layer was no longer a threat to you. In theory, the rocket nose cone heat shield design and jettison sequence makes sense. Nonetheless, while investigating the Apollo 11 Mission Rocket Launch, in an attempt to uncover NASA's method of deception, the launch sequence was not what was to be expected.

The next set of images illustrates the procedure for the jettison of the Nose cone and the space docking required before the spacecraft could have traveled to the moon.

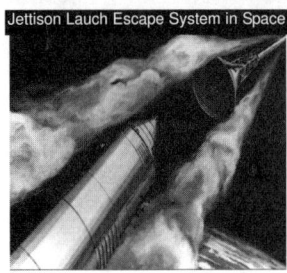 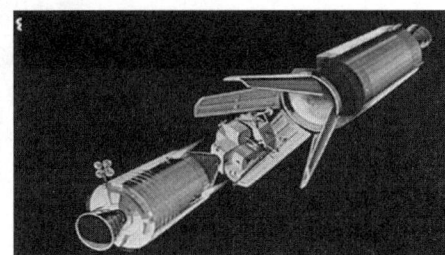  

Description Of the Apollo 11 Launch Sequence

While watching the Apollo 11 Astronauts soaring on their way to outer space at approximately 38 miles out to sea, shortly after the 1st stage fuel tank and o-ring was jettisoned, a few seconds later the nose cone heat shield suddenly was separated from the rocket. When the nose cone separated from the booster rocket prematurely inside earth's atmosphere, my first reaction was that there had been some type of terrible explosion during the Apollo 11 launch. However, the space rocket never slowed down. It just kept on going. Once the nose cone had reached a safe distance away from the rocket, the 2nd stage fuel tank ignited, and the rocket traveled on its way to outer space. Certainly, if the astronauts were still on the Apollo 11 rocket after the nose-cone heat shield was jettisoned, they would have burned to a crisp as they passed through the earth's ozone layer to reach outer space. This Apollo 11 film footage revealed NASA engineers made some type of alteration to the rocket. To find the answer, NASA's official rocket specifications need to be investigated even more thoroughly.

NASA Developed An Elaborate Launch Escape System For the Apollo Astronauts

When comparing both the Mercury and Apollo Saturn Rockets to the original Gemini rockets, a startling discovery was made. The mystery of what really happened to the Apollo astronauts, during their alleged trips to the moon, finally had a scientific answer that makes sense.

NASA's rocket design specifications revealed that the rocket's nose cone section was converted into an elaborate launch escape vehicle and no longer served as a heat shield.

The customization to the rocket's nose cone provided a method for NASA to remove the astronauts from the booster rocket while it raced towards outer space at over 6,000 miles per hour without being detected by the people watching the event from the ground. This explains, while until now, no one has been able to prove how NASA fooled everyone into believing that they launched the astronauts into outer space.

LAUNCH ESCAPE VEHICLE CONFIGURATION

Astronauts separate from rocket and drop back to earth during Launch

The Launch Escape Vehicle

Here is how NASA's launch escape system worked. As everyone focused there eyes on the spectacular view of the rocket traveling towards outer space, no one realized the astronauts were actually jettisoned from the rocket over the Atlantic Ocean and returned to earth hidden inside the rocket nose cone section. That's right. According to NASA's own records, the nose cone of the Apollo rockets were modified to encapsulate the command module that the astronauts were riding in and allowed them to return safely to earth.

The customized rocket nose cone was referred to the as "The Launch Escape Vehicle" and the "Launch Escape Tower" by NASA engineers. At the time this book was written, the Launch Escape Vehicle specifications were available in NASA's archives, and these information was obtained from NASA's web site.

To return the astronauts safely to earth, while remaining seated inside the command module that was hidden inside the shell of the Rocket Nose Cone. The Nose Cone was now an elaborate Launch Escape Vehicle customized with special landing equipment. Once back on earth, the astronauts staged the remaining portions of the moon landing in one of NASA's top secret moon hoax movie studios.

Once again, there is no real mystery in how NASA faked the moon landings. It just takes a little bit of thinking and some common sense to see right through the hoax. After making the discovery of how the astronauts were safely removed from the booster rocket, before going into outer space, these next pictures now make sense.

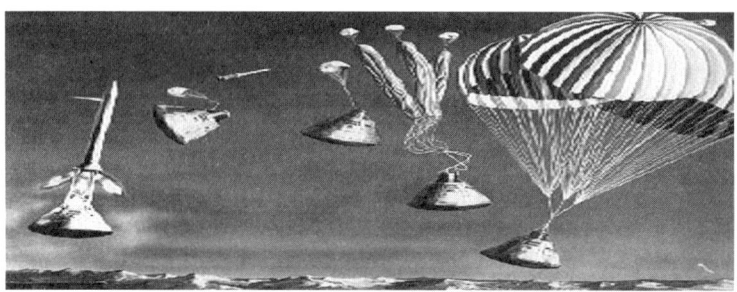

Confirmation NASA removed the astronauts from the spacecraft using the Launch Escape System.

An object discovered in a chaotic Earth orbit on September 3, 2002 is believed to be the third-stage rocket and service module of a Saturn 5 rocket used for the Apollo 12 Mission, according to a scientist at NASA's Jet Propulsion Laboratory. Did NASA forget that this section of the spacecraft was supposed to provide the engine power to return the astronauts back from the moon? Or is NASA purposely leaving everyone another clue about their moon-landing hoax, in hopes of putting an end to this ridiculous nonsense.

The presence of the section of the spacecraft would only be expected to be found, if the astronauts were ejected from the rocket using the Launch Escape system and remained on earth during their alleged missions. Because, after the astronauts were to have returned from the moon and commence the re-entry command module separation sequence, the rocket's 3rd stage section was to have quickly burned up in earth's atmosphere. It was not supposed to be drifting in an orbit originating near the moon.

Orbiting Pattern Of Recently Discovered Unmanned Apollo Spacecraft Helps Prove Moon Landings Were Faked

Based on the object's orbiting pattern it suggests the spacecraft stalled at approximately the same distance as the moon is from earth and started it's chaotic (irregular) orbit around the earth. This indicates NASA simply sent the unmanned rocket toward the moon and left it there. The diagram below shows the unmanned Apollo spacecraft's orbiting pattern to help illustrate how it's original orbiting point starts at the same distance the moon is from earth.

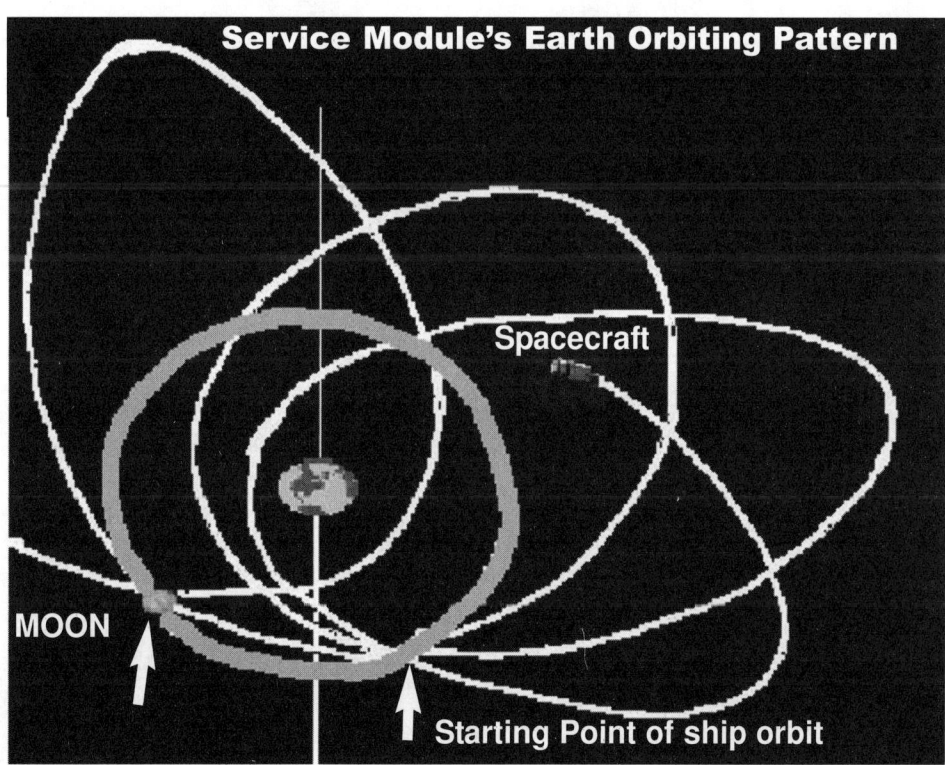

An Illustration of How The Launch Escape System Worked

The next illustration is designed to show how the astronauts escaped from the rocket as it raced over 6,000 miles per hour towards outer space. The unusual looking parachute was used to help camouflage the astronauts from the viewers, once the command module was separated from the booster rocket over the ocean.

Apollo Astronauts Escape Over Ocean

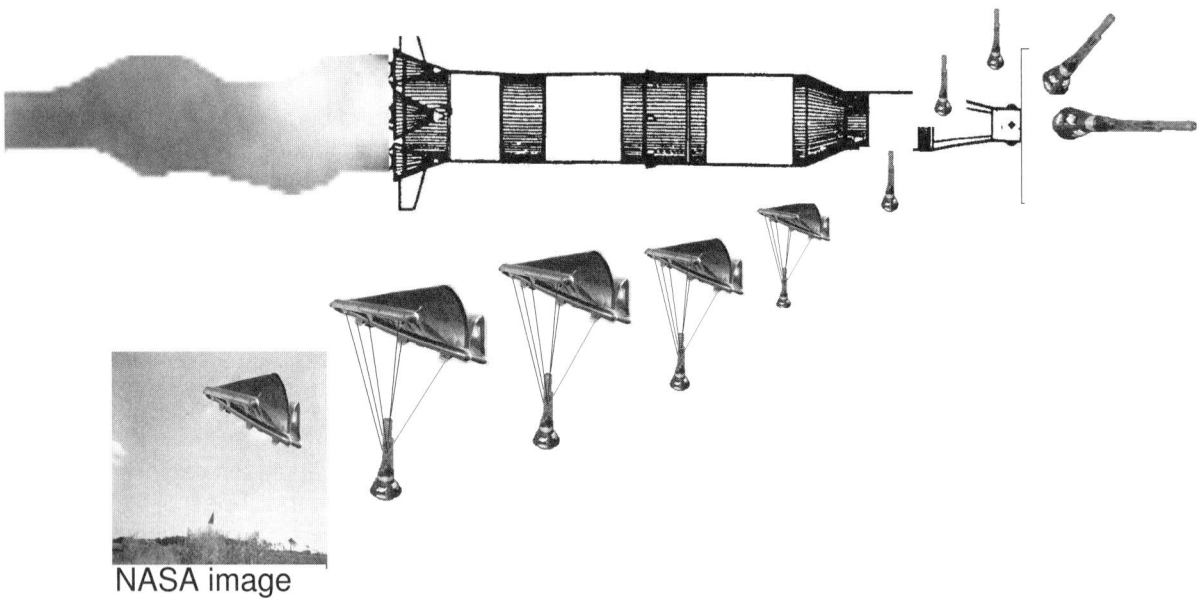

NASA image

You Can Bet These People Know How the Escape System Worked

As the Apollo Astronauts removed the Launch Escape Vehicle from the booster rocket, look at the way they were giggling to themselves. You would too, realizing your were not going to be launched on the rocket into deep space and vaporized by cosmic radiation.

NASA ID# 10073465

Chapter 7 - Part 3

The Astronauts' Space Flights Were Also Staged Here On Earth

The Astronauts' Space Flight Was Staged Here On Earth

After determining how NASA faked the astronaut's trips to the moon, it was really not that difficult to confirm how NASA faked the Apollo space launches with the Launch Escape Vehicle. It was obvious they had to use an earth based simulator. The graphics below show the same Apollo astronauts, who claimed to go to the moon, climbing into a customized mission command module simulator at the Kennedy Space Center.

10075597

10074864

10074865

ap17-KSC-72PC-386

While searching through the various flight simulator photos, a remarkable discovery was made. There was a message left by astronauts, Jim and Buzz, possibly a clue for people of the future to find.

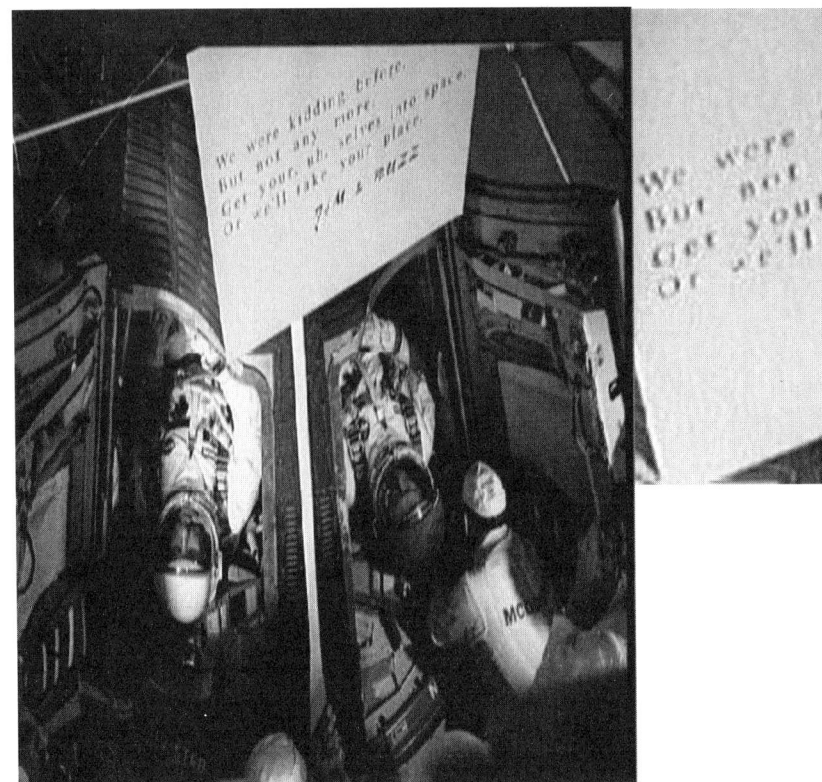
NASA ID# 10074365

The message read: "We were kidding before. But not any more. Get your, uh, selves into space. Or we'll take your place." JIM and BUZZ

Space Walk Staged On Earth

The question that needs to be answered is: were the astronauts simply conducting mission training photo shoots, or were they actually filming the fake moon landing scenes? There is clearly a very fine line of what was the true intent of taking these photos. Were these space walk photos used in some fashion to produce realistic looking space photography, or were they only for training purposes?

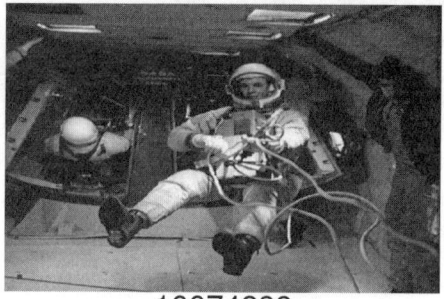
10074288

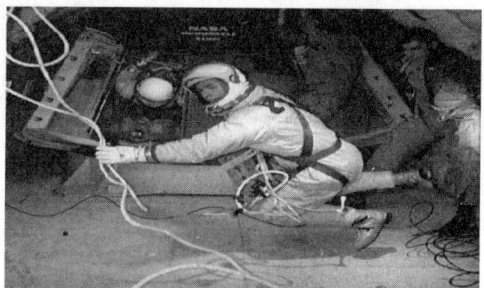
10074287

The only way to find the answer was to see if we could find simulator photos that were identical matches to the photos alleged to have been taken on the moon. After analyzing many of the space walk training simulator pictures, we discovered many matching photos alleged to have been taken on the moon. Once again, proving the moon landings were faked. To produce the space walk photography, NASA used a combination of at least three methods.

1. Airplane Simulators Were Used To Create Space Walk Photography.

The following sets of images were obtained from NASA's archives, and are perfect examples of how the Apollo training simulators were used to produce the moon landing hoax pictures.

Airplane Training   **Outer Space Picture**

Training

10074510

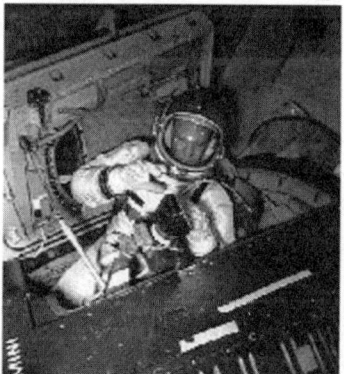
Training

10074564

2. Underwater Tanks Known As Neutral Buoyancy Simulators Were Also Used To Produce Space Walk Photography.

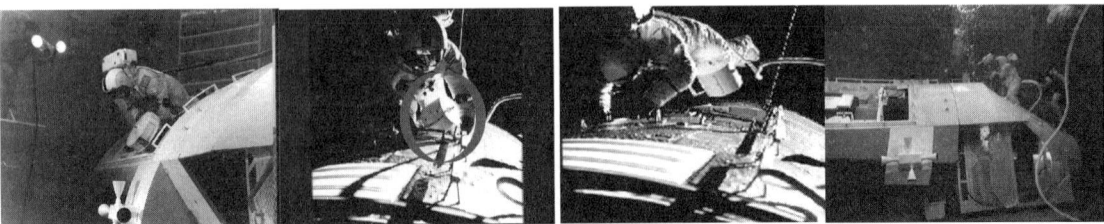

3. Combination Of Airplanes, Water Tank Simulators, And Manikins Were Used To Create Apollo Mission Photos.

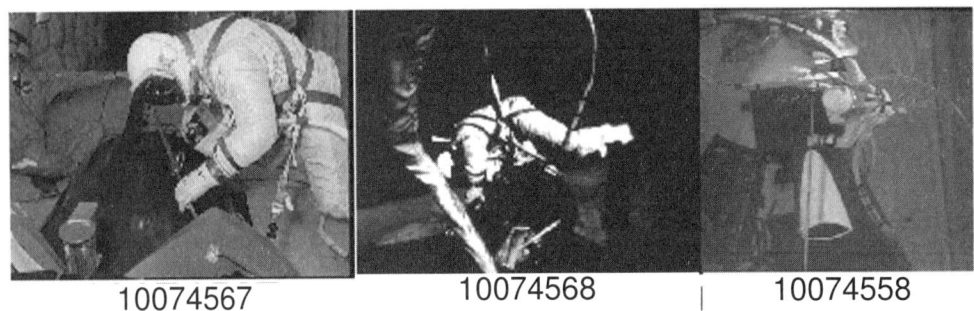

10074567   10074568   10074558

CHAPTER 7 - Part 4

How the Splashdowns were faked

How Return To Earth Splashdowns Were Faked

One of the best clues that NASA never sent the astronauts (actors) into orbit is the photos taken of them returning to earth in the capsules. Below are just a few of the many illustrations taken by NASA photographers as they witnessed the alleged arrivals (splashdowns) of the Apollo Spacecraft. What a historical event it must have been. NASA was exceeding their own wildest expectations of sending men to the moon and returning them safely.

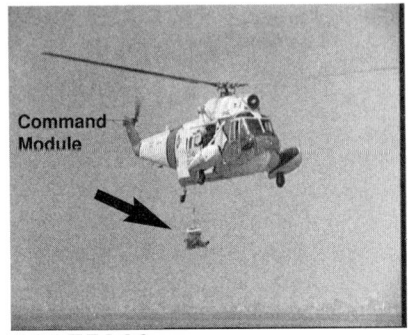
10075369

GPN-2000-000994

Let's think this through a little. How could NASA photographers have been waiting to film every splashdown? NASA would have only been able to pinpoint the landing sites within approximately an 80-kilometer (50-mile) area. With such a large area to cover, it would have been impossible for NASA to get a ship close enough to have taken any of these pictures. If that's the case, then why do we see helicopters hovering above the landing site before it even hits the water? Furthermore, how can there be rescue ships and even a rubber life raft waiting in the water not far from where the capsule hits the water? Is that even possible? Not according to other splashdown recovery attempts. For example, shortly after the Columbia tragedy grounded NASA's shuttle fleet in 2003, the International Space Station depended on Russian craft. The United States and Russia worked together to return several astronauts from the International Space Stations.

In MOSCOW(Reuters), on Monday, May 5, 2003, Sergei Gorbunov, spokesman for Russia's Rosaviakosmos, said "The task of the rescuers is to find the crew within four hours. They found them in two hours and 20 minutes. So, while this was not quite standard, it is within the norm (normal time frame)."

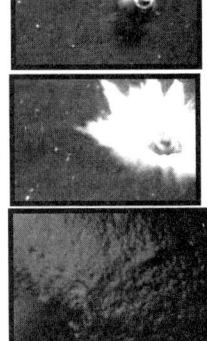
Apollo 15 Film crew with a view feet of splash down

Sergei Gorbunov also said, capsules returning from the space stations have landed some 3,200 km off target. He continued by saying that one crew had to wait three days for rescuers to cut through thick ural forests to find them. Sergei Puzanov, a NASA spokesman, said, "It's not normal, but it's not that unusual either, let's put it that way. "

This means some thirty-years later, using the finest modern day technology, it still took 2 hours and 20 minutes to find a splashdown location. Again, with primitive tracking capability, back in the 1960's and 1970's, how could the American photographers and rescue teams have been waiting at the exact splashdown location? The splashdowns simply could not have been real.

The answer is these Apollo splashdown sightings were simply training mission films created by NASA and presented to the world as actual mission film footage. When you compare the flight profile of a typical Apollo Spacecraft Telemetry, as it travels to and from the moon, this is evident. A diagram is to the right.

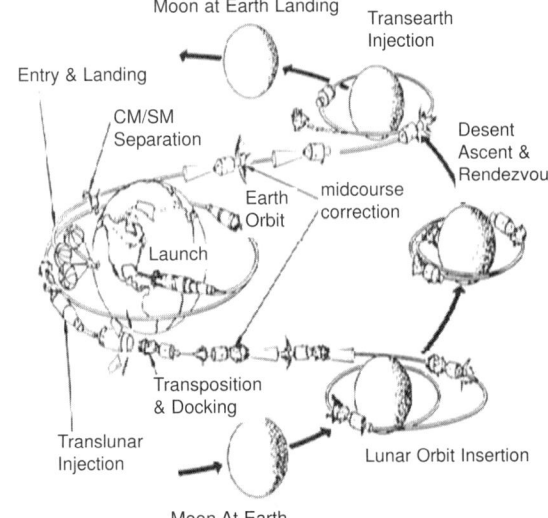
Lunar Flight Profile

7-26

Astronaut Training Missions Were Used To Fake Spacecraft Splashdowns

NASA simply transported the astronauts and capsules on a ship out to sea. Subsequently, they dropped the astronauts and capsules from a helicopter just a few hundred feet from where the recovery crew and photographers were waiting. The next mission training photos reveal a helicopter transporting the Apollo Space Capsules, in a net, to be dropped right where the movie crews and rescue workers were waiting.

EL-2002-00428   NASA ID# 10075369

There is plenty more evidence which proves the astronauts conducted splashdown training missions. This is illustrated in the pictures below. It would have been very easy for NASA to pass these astronaut splashdown trainings off as the real thing. This type of movie film work was common in almost every action movie.

Below are pictures of the Apollo 11 mission splashdown. The portrait, on the left, displays the Apollo 11 Astronauts being transported on a boat while they were standing next to the space capsule. In the middle picture, the astronauts were observed climbing into the capsule. The above picture, on the right, portrays the rescue workers opening the door to help the astronauts out, and depicts the harness used to lift the capsule when the helicopter was still attached.

If NASA was conducting splashdown training missions, they could have easily faked all the Apollo mission splashdowns as well. After an artist and a photographer did a little touch-up work, the two rescue scenes below were identical.

NASA # 10075368   NASA # ap11-s69-34881   NASA ID# 10075304

How NASA Faked the Mission Equipment Capabilities

Chapter 7 - Part 5

The Lunar Rover Vehicle

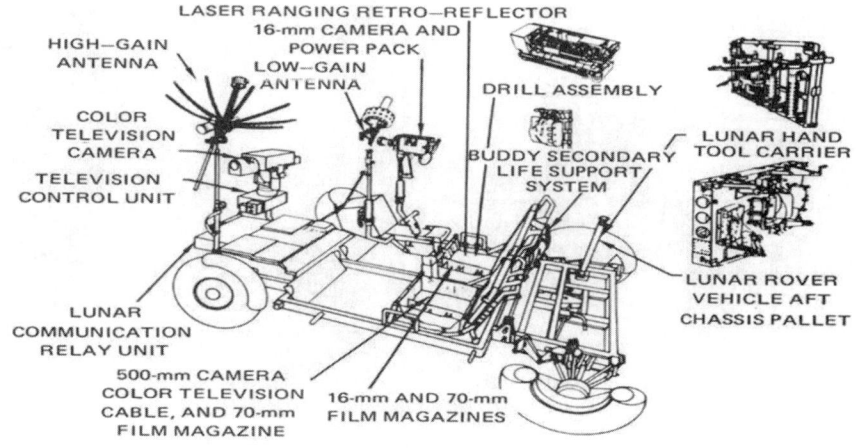

7-29

How NASA Faked the Mission Equipment Capabilities

The History Of the Lunar Rover Vehicle

NASA first moon rover built in the late 1960's

In the summer of 1969, NASA Commander and Chief, Wernher von Braun, took the first Moon Rover Vehicle prototype for a test spin at the Marshell Space Center. Wernher von Braun states "As there are no superhighways on the Moon (yet), all vehicles must have cross-country capability".

Jerry Zulawski, a science fiction writer, wrote about a moon rover vehicle in 1901. However, Wernher von Braun, in the 1950's, published a story in Collier's Magazine of his vision of a 10-ton tractor-trailer being used on the Moon. This explained the excitement on his face when he got the opportunity to ride this new prototype NASA's engineers have developed. It was exactly what Mr. Von Braun had envisioned. What was he thinking?

Below are a few of the additional rover suggestions other people made, which NASA engineers quickly focused their attention on.

A fulliscale mock uo of the Grumman Mobile Laboratory

A proposed concept for a lunar surface vehicle by Grumman

The Lunar Rover Vehicle Capabilities Were Also Exaggerated

Based on the technical specifications of the Lunar Rover Vehicle outline, in NASA's Lunar Rover Vehicle Operations Handbook #LS006-002-2h, there were many design flaws that clearly would have made it impossible for the Rover Vehicle to have ever mechanically functioned on the moon. This is not a hypothesis; it's a fact! According to NASA's official design specifications, there was absolutely no way the Rover could have functioned on the moon.

Also, this is not my opinion. These findings were based on basic electrical and mechanical principles. Again, I personally have over 6 years of extensive University Training on the subject and over 25 years of experience. Although this type of education is not necessary, with decades of scientific advancement and a much greater understanding of the moon itself, it's now very easy to determine the Rover Vehicle's flaws and prove it could not have worked on the moon.

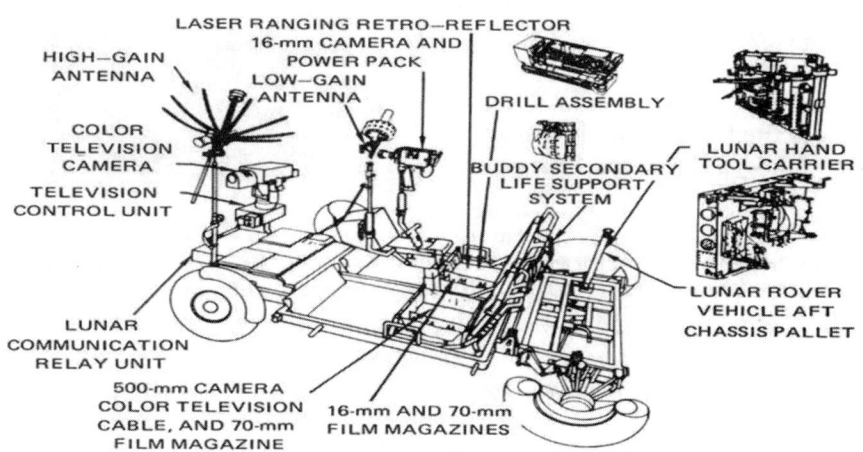

The Lunar Rover Vehicle Was Incapable Of Ever Starting Under the Conditions Found On the Moon's Surface.

NASA Used Incapable Thermal Sensors And Pressure Relief Valves

Based on the design specifications found in NASA's Lunar Rover Vehicle Operations Handbook #LS006-002-2h, the primitive electrical design of the Lunar Rover prevented the vehicle from even starting in outer space in the first place. There were many thermal sensors and pressure relief valves incorporated into the design of the vehicle with pre-determined shut down thresholds that rendered (made) the vehicle useless in outer space.

For example, according to NASA's Lunar Rover Vehicle Operations Handbook #LS006-002-2h, the Lunar Rover drive motor included a gearbox thermal switch that shut down the system when the temperature reached 200ºF. Since the temperature, in the sunlight on the moon surface, was a constant 250ºF, according to NASA's published design specifications, the Rover Vehicle was incapable of starting on the moon. According to notes made by NASA's engineers, in the Lunar Rover Vehicle Operations Handbook, these thermal switches were a major problem during the testing here on earth in the deserts, which had a much lower ambient temperature. The Rover would constantly overheat, shut down, and quit working until it cooled off.

NASA Used A Battery System Incompatible With the Conditions Found On the Moon Surface

According to NASA's Lunar Rover Vehicle Operations Handbook #LS006-002-2h, the Lunar Rover battery did not function below -20ºF or above +120ºF. That's right. The Lunar Rover battery system did not function below -20ºF or above +120ºF. Since the temperature on the moon's surface was approximately +250ºF, in the sunlight, and -250ºF in the dark, NASA's own records indicate that the Lunar Rover Vehicle wouldn't have started on the moon's surface.

The Rover Vehicle's Traveling Distance Was Greatly Exaggerated By NASA's Movie Producers

NASA's Engineers Obviously Did Not Communicate Well With the NASA Moon Landing Movie Producers. According to NASA's engineers' design calculations, in the Lunar Rover Vehicle Operations Handbook #LS006-002-2h, the Rover Vehicle had nowhere near the capability to conduct the Apollo Mission activities claimed to have taken place during the alleged Apollo Missions. NASA's movie producers extended the capability of the driving duration of the Rover Vehicle as much as 3.12 times beyond its limits. We're not talking about a little exaggeration here. It was a whopping 3.12 times beyond the Rover Vehicle's limitations. By referencing NASA's Lunar Rover Vehicle's Operations Handbook, #LS006-002-2h, and making a few simple calculations on the capabilities of the Rover Vehicle battery system, this can easily be proven.

How To Calculate the Rover Vehicle's Travel Distance Using NASA's Recorded Design Specifications.

Based on the technical specifications of the rover, in the Lunar Rover Vehicle Operations Handbook #LS006-002-2h, the Lunar Rover operates on two low powered primary zinc batteries. According to this official NASA specification, the Rover was equipped with two 36 Voltage DC silver zinc batteries as its fuel cell with a capacity of just 115-ampere hours. With the approximate payload of 800-pounds, the small amount of combined power of these two batteries would drain the power supply very quickly.

Why would NASA even send such a vehicle with inferior capabilities to the moon with the astronauts anyways? According to their own design specifications, the rover vehicle was basically worthless. Since there was no means of recharging the Rover Vehicle batteries fuel cell, once it left earth. One thing was for sure, NASA's claim of how much the astronauts used the moon rover during the Apollo missions were extremely exaggerated.

Obtaining NASA's design specifications, on the moon rover, provided another excellent opportunity to prove the moon landings were filmed here on earth. Anyone with some basic math skills and NASA's Lunar Rover Vehicle Operations Handbook #LS006-002-2h can verify that the rover was incapable of traveling anywhere near the distance NASA claimed before its power supply would have run out.

Formula To Prove Lunar Rover Vehicle Scenes Were Filmed On Earth.

According to NASA's Lunar Rover Vehicle Operations Handbook, #LS006-002-2h, the Rover only had a maximum drive time of 63 minutes. Nevertheless NASA claimed, during the Apollo 16 mission, which was a 3 day mission on the rover vehicle, that the vehicle went an incredible 197 minutes without being recharged. That equals over 3.12 times more than it was capable of traveling. Can you imagine how much money everyone could save if NASA really had such a magic car? If your car normally gets 400 miles to tank, you could now travel over 1,200 miles without filling up again. Even after the rover vehicle somehow managed to operate over 3 times longer than it was capable of doing, it still had enough power to continue running the on-board camera after the astronauts left, which was simply impossible.

Let's break this down further, and see if we can scientifically prove that the rover could not have functioned as NASA claimed. With a maximum of 63 minutes drive time on a smooth level surface, how far could the rover have traveled? Based on the overall design of the rover, it would have required at least 19 of the 63 minutes as rest time. This was a standing time where the rover vehicle was not moving. This is like a cool down period. This would have left no more than 44 minutes for driving time.

With a maximum traveling drive time of 44 minutes, and the approximate payload of 800 pounds, at 10 miles per hour, the most the rover vehicle could have traveled, on the Moon, would have been 7.4 miles. This was hardly enough vehicle fuel capacity time to have completed any of the Apollo Moon Landing Missions NASA claimed to have performed.

If NASA had such superior technology, back in the 1970's to build an electrical car with such efficiency, than by those standards, we should not have seen electrical cars that can travel over 80-110 miles on a single charge, but more like 330 miles without a recharge. Today, we don't. Why would NASA be hiding this secret technology from the public? Is NASA hiding this secret from the public out of fear that some other country might get their hands on it? Are they scared of some great secret weapon, such as an electric car that fights air pollution?

Of course not, the logical explanation is that the Lunar Rover Vehicle was designed simply to make NASA's Moon landing hoax as entertaining as possible. Realizing a whole new understanding of what we should be looking for, when looking at the Moon rover on the moon, we should be searching for things the astronauts would have experienced if the filming was done here on Earth.

After going on the assumption that the rover was actually driving in a desert somewhere, rather than the Moon, it only took a few minutes to confirm this assumption. Below is some solid evidence that the astronauts were not driving the rover on the Moon. For instance, the presence of "Desert tumbleweed" getting hung up on the rover vehicle. Tumbleweeds do not grow on the Moon, but it is abundant in the deserts of North America.

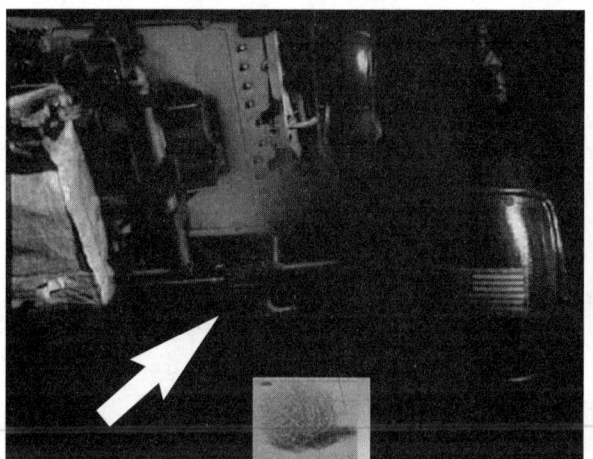
"Desert Tumbleweed" found on moon

Apollo 15 Astronauts Jim Irwin ans Dave Scott

Another Earth and Moon Scenery Match.

Once the Tumbleweed was discovered hanging from the Moon rover vehicle, it was easy to find the matching location on Earth where the filming of the Moon rover actually took place. It's funny how the hill, in the background of this rover testing facility in Arizona, matched the hill on the Moon where the astronauts were also testing the rover vehicle on Earth. Notice the astronaut is in his space suit. Was NASA leaving these clues in hope that someday people of the future would figure out the Moon landings were a hoax. Maybe NASA was afraid that, in the future, people in the American Government, would forget to tell everyone the Moon landings were faked. Maybe they were supposed to have told everyone and already forgot to. Only time will tell.

Three More WhizKid Experiments To Prove Moon Landings Were A Hoax

To decide whether or not the astronauts traveled to the moon, these first two WhizKid experiments could be conducted immediately.

Whizkid, Nicholas (age 15), came up with an experiment that could be done for free; it was also very simple and would only require getting NASA's consent to complete the project. Nicholas was a very valuable member to the "WhizKids Team" because he is gifted with some kind of photographic memory. Nicholas seems to know facts about many different subjects. If you were to ask Nicholas, for instance, what song was playing on the radio at any given time, he would not only tell you the name of the song, but also the history behind the song, including what the band's name was, and the names of the band members.

Nicholas last week, decided he was going to take some evening college courses at the local community college. He found that high school was not challenging enough by itself. When he went to sign up, they asked him to take an assessment test. The college wanted to know if he was capable of attending the school at an early age.

After Nicholas finished the test, he was asked to wait a minute by the registration desk. Suddenly, two people of the college staff approached him and escorted him to the administration office. In the office they asked Nicholas if they could see his cell phone. Nicholas said, " I don't have a cell phone." They said, "Well, if you don't have a cell phone, what were you using to cheat on the test."

Nicholas told them he did not cheat on the test and asked them why they thought he did. They told Nicholas he scored an A+ level on the test. This was not normal for someone his age to achieve on a college assessment test. They then said, "Well, if you did not cheat on the test, you must be autistic or something because you not only got an A+ on the college test, but also it was a 2 hour test. Nicholas finished it in just a little over 15 minutes, which makes no sense. Nicholas agreed to take the test again and scored higher. Nicholas stated that one day he wants to become a politician.

WhizKids Experiment #2:

Here is Nick's experiment. The WhizKids would like to perform a very comprehensive radiation and meteorite bombardment test on the Apollo space capsules and compare those measured against one of the remaining NASA space shuttles. Nicholas believed that, since the space shuttle never ventured outside the safety of earth's orbit into the deadly radiation of deep space, the radiation levels of the shuttle should be much less than the Apollo capsules, which had been exposed to the extreme levels of radiation.

ap15-s71-42037

Whiz Kids Experiment #3:

Whiz Kid, Tony (age 10), says he wants to be a special effect movie producer and actor for NASA just like Wernher Von Braun. Tony's experiment could be conducted almost immediately. The Whiz Kids would rent the Hubble Telescope, from NASA, and search the moon's surface to obtain close-up pictures of the equipment supposedly left behind by the Apollo Astronauts.

WhizKid, Mitchell (age 14), believes it would be almost impossible to actually view the objects with the HST (Hubble Space Telescope). Nevertheless, he thinks recording light energy variance, in the moon's reflection, could easily detect the larger objects allegedly left behind.

Mitchell suggests that, if NASA would let them temporarily put the Hubble Space Telescope into "waylay mode", at a predetermined position with recalibrating the HST selectivity and sensitivity to maximize the resolution efficiency, then they could point the telescope to a predetermined place where the moon will be moving into view, and using the fast exposure mode, they could capture the images.

Sergey, and the other Internet WhizKids from Russia, confirmed Mitchell's theory. Agreeing that the magnification capabilities of the Hubble Telescope are most likely too primitive to actually see the objects. They calculated, with an earth-moon distance of 239,000 miles, the HST's resolving power of 0.05 inches. Hence, the smallest object that could be seen would be approximately 300 feet. Since the largest dimension of any hardware supposedly left behind on the moon is thirty-one feet, which is the diagonal distance across the LM's footpad, it could not be seen with the Hubble Space Telescope.

Although they are also convinced by properly sorting the individual color spectra, and by breaking the ultraviolet and infrared light images, the objects on the moon's surface would easily be detected while still protecting the Hubble Space Telescope from light sensitivity damage.

WhizKids Experiment #4:

To further confirm the theory that the Astronauts were jettisoned from the Saturn Rocket, before reaching outer space, Whiz Kid, Alex (age 9), suggested that the Hubble telescope be used to scan earth's orbit. If NASA did not go to the moon more Apollo Lunar Module debris could be identified in earth's orbit. We should still be able to detect objects like the 3rd stage spacecraft drifting around in earth's orbit. WhizKid, Andrew (age 7), suggested that if any Apollo equipment was found in earth's orbit, it should be mapped out, or even collected during space shuttle missions, so other spacecrafts do not hit these Apollo objects when re-entering earth's atmosphere.

The WhizKids have set up a fundraiser to rent the Hubble Telescope for this experiment. If you would like to help the WhizKids reach this lofty goal, you can make a donation at www.whizkids.tv or moonbloopers.com.

CHAPTER 8

NASA's Moon Landing Studios

Chapter 8
NASA's Moon Landing Movie Studios

After carefully analyzing all the evidence provided by NASA, there were many clues revealing how the astronaut's training simulators were used to stage the moon landing missions. Yes, there is an abundance of evidence that NASA staged the Moon landings here on earth. Finally, we can put an end to NASA's false claims.

There are many things wrong with NASA's claims that make no scientific sense. For this reason it's almost unimaginable they have gone undetected for so long. No more trying to explain why the flags were seen blowing in the wind on the Moon surface. There is no wind on the Moon! Discussions about why the Apollo astronauts did not notice any stars in outer space can stop. If there is very little gravity on the Moon, then why did the Moon dust, kicked up by the astronauts and Moon rover, immediately fall back to the surface? The list goes on and on! NASA's claims of landing on the Moon simply does not make sense.

No Blast Craters

One of the most obvious clues to the Moon landing being a hoax was the lack of blast craters under the Lunar Module Spacecraft. This is a major contradiction to what would be expected and is very helpful in answering where and how NASA constructed the Moon landing sites here on Earth.

Painting of final decent

At first, almost everyone found it strange that there wasn't a blast crater under the spacecraft in any of the pictures alleged to have been taken on the Moon's surface. Why in the Apollo films was the spacecraft (Lunar Lander) seen descending on to the moon's surface, while only a small amount of dust was blown around. This is just another major oversight on NASA's part that can scientifically prove the Moon landings were faked on Earth.

If the spacecraft with a Grumman rocket engine were running during the descent onto the Moon, it would have been blowing out dense exhaust gases from the engine. It would have burned hypergolic fuels with roughly 3,000 lbs of exhaust thrust and a fifty-four inch diameter. During touchdown, it should have blown up tons of dust and created a hole so big the spacecraft would have fallen into it. Furthermore, according to NASA's own records, the heat produced by the rocket engine was over 1,800 degrees. The exhaust heat alone from the engine would have cooked the astronauts.

The NASA film footage provides another piece of evidence that supports the theory that the Spacecraft engines were not running. Every alleged Apollo LM Spacecraft descent on to the Moon surface showed just a very small amount of dust being blown around, which is nowhere near what would have been expected. This suggests that, under the spacecraft, a large fan was attached. The fan was lowered from the crane onto the artificial Moon surface at the Langley Space Center.

This also explains why the photos show the Moon surface beneath the spacecraft either undisturbed, or containing astronaut's footprints in the moon dust. If the Apollo Spacecraft engines were running, try and picture this condition, it could not have occurred on the Moon surface.

All the Apollo Astronauts Should Have Lost Their Hearing

If these films were actually taken on the moon, then, while descending onto the moon surface, why was there never any engine noise heard? Because the microphones, used by the astronauts, were located inside their space suits and designed to pick up sound in their immediate vicinity, the noise would have echoed and been extremely loud.

Since the astronauts were supposedly riding on top of the rocket engines, the noise from the high velocity exhaust jet should have made them deaf, if not killed them. These astronauts were supposedly right on top of the engine; this was not like a Space Shuttle, or airplane cockpit where the pilots are over one hundred feet away from the engines.

A Simple Explanation Why the LM Spacecraft Flight Was Faked:

Some of the teenage kids were a little confused with all the laws of gas expansion and particle collision in a vacuum. Therefore, they asked me, is there a simpler way to prove the spacecraft landings were faked? The kids were right. Sometimes it's easy to make things more complicated than they really are, so here is a simple explanation.

NASA claimed the same Lander Spacecraft design was used on all Apollo Missions. However, when the Lunar Rover Vehicle was added to the spacecraft during the last few missions, there appeared to be no changes made to the design specifications, which could have compensated for the extra weight. The addition of the Lunar Rover Vehicle would have shifted the center of mass to one side of the spacecraft. This would have thrown the lander off balance. Thus, the spacecraft would have spun out of control and crashed onto the Moon surface.

The addition of the Lunar Rover Vehicle would have shifted the center of mass to one side of the spacecraft, throwing the Lunar Lander off balance; thus, causing the Lander Spacecraft to spin out of control and crash into the moon surface.

Astronauts Should All Be Dead

If that's not enough, in the Advanced Mathematics Chapter of the book, you can read how the Apollo Lander Spacecraft design prevented it from ever functioning in outer space altogether.

The LM Spacecraft Weighed Approximately 17 Tons

Another strange condition that puzzled everyone was the weight of the spacecraft resting on the dusty surface of the moon. The photos below were supposedly taken during the Apollo 11 Mission on the moon. The picture on the left shows there was no blast crater under the Lander Spacecraft. The picture in the center shows one of the LM spacecraft landing pads resting on top of the moon dust surface and barely making an impression in the soil. If the spacecraft weighed 17 tons, how come it did not sink into the powdery moon dust surface. Nevertheless, there were astronaut footprints all over the place.

This condition defies the laws of physics. Are we to believe that heavier objects become lighter on the moon surface? Could there be some type of deception on NASA's part here?

The picture on the right is Astronaut Buzz Aldrin setting up the Moon testing experiment. You clearly see the imprint left by his foot. Further, compare his footprints with the LM Landing Pad picture. If the LM Spacecraft was considerably heavier than Aldrin, why was it not submerged in the soil?

The reason the leg is not submerged in the moondust is because it is mounted on stage platform pedestals. As shown in the next set of pictures, the soil was later brought in and filled up to the bottom of these legs.

LM Spacecraft Was Placed On Landing Pad Pedestals Then the Artificial Moondust Was Added

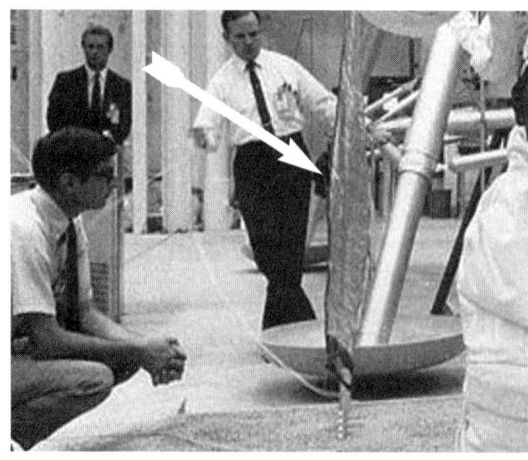

Examples Of Moondust Added To Camouflage the Landing Footpads.

This next set of LM Spacecraft landing site photos shows the landing pad with moondust built up over them. Undoubtedly, if there was moondust this loose around the landing site, during the landing decent, an engine thrust of 3,000 psi would have moved the dust hundreds of feet away from the landing site.

10100415

10100436

10100446

Evidence Showing How Plaster And Moondust Was Built Up Around the Stage Platform Pedestals.

To avoid the stage platform pedestals being detected in the photography, the pedestals were wrapped with plaster. Next, they were surrounded with a thicker moondust mixture and a final thin layer of moondust. Below is an example of how NASA constructed the moon surface around one of the Apollo 11 LM spacecraft legs.

as11-40-5920

as11-40-5926

Examples Of Moon Surface Construction Outside Kennedy Space Center

In the next set of pictures, you see precisely how NASA formed the Moon surface for the Apollo movies with the pedestals visible. To determine what methods NASA was using, we needed to keep in mind that each Apollo moon landing mission must have been filmed at one of NASA's elaborate movie studios. After that, using a time lapse photo system, it was easy to determine how NASA built each Moon landing site. To illustrate, look at the snapshots shown below.

Background Painting Starts Here

20135237

For a lot of the moon scenes, NASA used a fairly flat stage platform with the LM Spacecraft landing pads mounted to pedestals.

The LM Spacecraft Never Left the Kennedy Space Center

These next set of graphics are another perfect example of how NASA constructed the Moon surface appearance. For this moon mission photography, the LM Spacecraft remained inside the Kennedy Space Center.

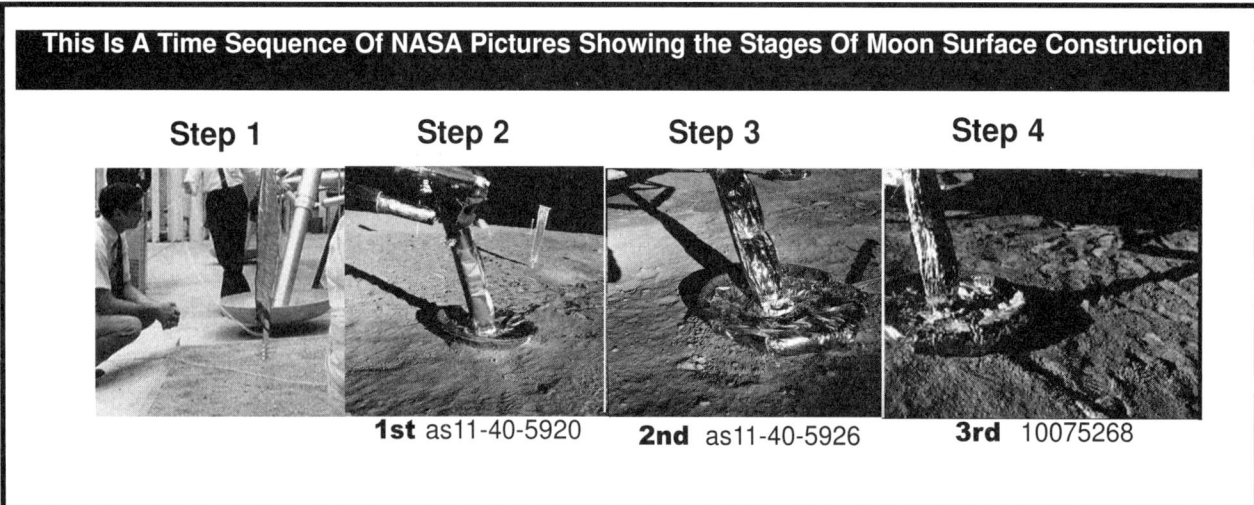

First, the LM Spacecraft was lowered onto the pedestals. Secondly, the artificial moondust mixture was added. Thirdly, the stage props. Finally, the pictures were taken. These pictures are additional proof of the extent NASA went through to stage the fake Moon landings. Keep in mind that this is one of the many examples of NASA's Apollo training simulators being used to replicate backgrounds.

Stage complete as11-40-5873

No Footprints, Or Tire Tracks Help Solve the Mystery

Although there were countless things that don't make sense, this next problem with the images is one of the most revealing. In many photos, various items, like the Moon Rover, were seen out of the reach of the astronauts. Additionally, there were no footprints, or tire tracks, leading to, or from, the items. Without a doubt, this was a powerful piece of evidence indicating the Moon landing sites were man made on a movie studio here on earth.

The following visual renderings are ideal examples of how the NASA stage crew set up each moon landing scene before the astronauts arrived for filming. When you look at this picture supposedly taken on the moon during an Apollo Mission, it's immediately obvious something is wrong. There are clearly no footprints, or tire tracks, leading to or from the Rover Vehicle. This shows the Moon soil was added after the Rover Vehicle was placed in that spot. The vehicle was placed here long before the astronauts arrived on the scene.

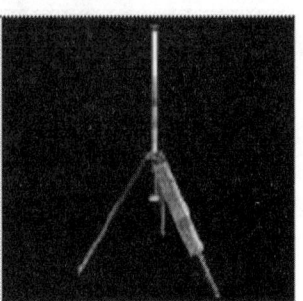

NASA ID# 20134558

After analyzing many pictures with similar conditions, in an attempt to find out what photographic trickery was used, there was a common link in all the photos discovered. NASA filmmakers used the "Gnomon Tripod Color Chart" as a stage marker. Gnomon Tripod let the astronauts know where to stop walking, and where their photo shoot was to take place. The astronauts were informed before advancing to the next section on the movie set. The Gnomon Tripod also helped the astronauts locate certain stage prop items that were camouflaged on the Moon surface.

In addition, the "Gnomon Tripod" provided significant clues as to how NASA was able to assemble all the different artificial Apollo Moon landing sceneries. It was actually a very simple step-by-step process they followed that could have easily been done with less then twenty or so workers.

In some cases the stage first sprayed a thin layer of artificial Moon dust mixture over the stage to cover their tracks. Then, crew members placed a "Gnomon Tripod" as a stage marker for the position of the next scene. Throughout the rest of this book, every aspect of the artificial Moon scenery will be broken down and explained in greater detail. However, it was impossible to cover it all at once. To fully understand NASA's method of deception, there are other related topics that need to be covered in other chapters. The rest of this chapter will cover some of the basic problems with the alleged Moon landing pictures as they relate to the stage construction.

Here are a few more visual renderings of missing footprints in the alleged Moon landing portraits. Again, the reason the stage props were missing was because the pictures were taken before the astronauts arrived on the movie stage.

NASA ID# AS12-47-6899

NASA ID# AS12-48-7069

Just imagine, it would have taken the stage crew several hours to set up each scene, so it was critical that the astronauts knew exactly where to step to avoid contaminating the next photo shoot area. To calibrate their cameras for lighting and shadows, the photographers filming the astronauts also used the Gnomon Color Chart.

Below are more pictures demonstrating that the "Gnomon Tripod" was always slightly ahead of the astronaut's position where they were working. Also, the moon surface, beyond the "Gnomon Tripod", was always untouched by the astronauts. This supports the theory that the astronauts were often told not to pass the "Gnomon Tripod" until after the photographers had finished their filming of a particular scene.

No Foot Prints Past the "Gnomon Movie Stage Marker"

To take this next illustration of the "Gnomon Stage Marker" three people were needed

The shot on the left shows the second astronaut holding a tool in his left hand while taking the picture. The portrait on the right displays the same astronaut standing to the right of the picture angle still holding the tool in his right hand. Since none of the cameras sent to the Moon were equipped with any delayed trigger capabilities, and since there were only two astronauts on the surface, how could this picture of the two astronauts have been taken?

How could three people filmed this scene with only two men on moon?

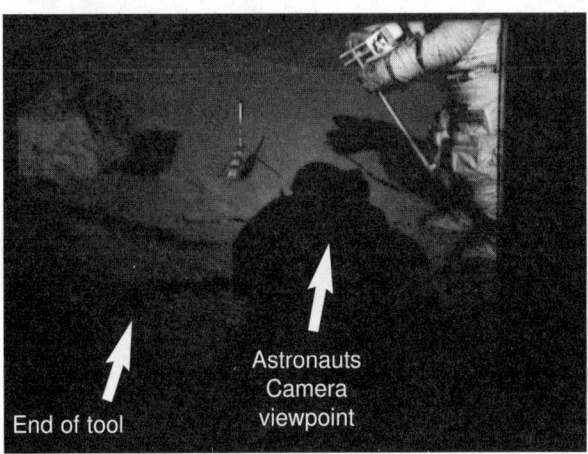

"Gnomon Stage Marker" Matched Earth Training Pictures With Moon Landing Hoax Photos

The rock formation below is an excellent example of how the same "Gnomon Tripod" was used on earth to film the astronauts during the alleged mission training. Notice how the Apollo training work area matches the scenery allegedly found on the Moon.

NASA ID# 20134558

It was obvious that the two images above were filmed at the same location. When you take the picture on the left and cover the rocks with some artificial Moon soil, and then take a photo from behind the rocks, you have an earth and moon picture match. In the earth picture, even the rock under the "Gnomon Tripod" was an identical match to that on the Moon photo.

NASA's Moon Landing Movie Studios

Below are samples of spouses watching their husbands fake the Moon landing at the Kennedy Space Center. We spotted the wives because no matter how hard NASA tried to make sure ever piece of evidence was covered up, they simple could not have thought of everything.

Women's Reflection In Astronaut's Helmets

Woman Reflection in helmet   NASA ID# as16-117-18852

Picture Of Kids And Shadows

ap17-KSC-72PC-379   Girl With Purse   NASA ID# 20147711   Boy

8-11

Movie Producer's Chair

Besides the fact that the lighting is totally incorrect in this picture, the size of the Earth in the horizon is six times too small and there are no stars to be seen anywhere. There is also a camouflaged movie producer chair sitting next to the LM spacecraft. This black-colored movie producer's chair did not even have to be special ordered it was one of the most popular colors during the Apollo era. It goes well with the movie set theme, don't you think?

Movie Producer's Chair

NASA ID# as17-151-23188

CHAPTER 9

Proof the Moon Landing Photography Is Fake

Chapter 9
Proof That All the Moon Landing Photography Is Fake

There are many items that make no sense in NASA's alleged Moon landing photography, such as strange lights in the sky, shadows traveling in different directions, and so on. After analyzing each of these problems, one by one, there is a logical explanation for every puzzle.

Each problem will be broken down and analyzed separately, but one of the simplest ways to prove all the Moon landing pictures were faked is to consider the environment the film would have been subjected to on the moon. Just the amount of heat and solar radiation exposed to the film would have destroyed it. In areas exposed to sunlight, the temperature on the moon reaches 250°F. In addition, the intense peaks of powerful cosmic radiation can be more lethal than a nuclear power plant meltdown. In spite of that, NASA's films were unaffected. On the moon surface, the film should have melted in the cameras.

Why doesn't the film show some fogging from the radiation levels? The tapes are in perfect condition. With the shielding NASA claims their cameras had, the heat alone would have curled up the film, and it should have been destroyed. Strikingly, even if the American Astronauts had actually gone to the moon, none of the pictures could have been from there.

Physicist, Dr. David Groves Ph. D., has performed radiation tests on similar films and found that a much lower radiation level (25 rem) applied to a portion of the film, after exposure, made the image almost entirely destroyed. For that reason, why didn't that happen to the Apollo Moon Landing films?

This theory coincides with an article in the English Daily Telegraph (April 26, 2002). Journalist, Robert Uhlig, tells us that while filming the space station, under construction at an altitude about 240 miles, "The astronauts had to race to shoot the film before it was fogged by the high levels of radiation in space." Toni Myers, the Earth-based director of this Imax 3D production, had this to say: "We had to get the film up to the space station, throw it across from the shuttle, shoot it, throw it back to the shuttle, and take it back. This all had to be done on the same flight to the space station, or it would have been ruined."

The WhizKid Andrew (age 8) also asked: "How could any of the pictures have come from the moon? The film would have never survived." Andrew said, "It sounds like taking some pictures, then putting it in your oven at 250 F for thirty minutes. After that, put it in a microwave oven for five minutes. The film would never be able to be developed, so how could NASA's film survive on the moon with similar conditions?"

To help provide the answer to the "strange conditions" that appeared in so many of the Apollo photos, each problem photo was examined. Many of these problems with the films are so obvious that, to realize the pictures are not from the moon and are fakes, individuals do not have to be a photographic expert. Not only is it easy to determine that the shoots are fake, but there is also overwhelming evidence that proves the photography was filmed on a movie stage here on Earth. Other problems found in the film simply cannot have existed on the moon. Again, no matter how big NASA's budget was, they can't change the laws of physics.

Examination Of the Apollo-Photo "Anomalies"

There are many visual signs on the photographs and films, suggesting they were not genuine. For example, most of the photo placements make sense. There are many lighting inconsistencies, and foreign objects that don't belong on the moon. Solid evidence shows all the film footage was prerecorded at multiple earth locations. Nevertheless, NASA still claimed the recordings were shot at the same time on the moon. How could this be? Since the still photography rarely matched the video footage. This is surely not possible, and later in this book, it will be explained in much greater detail. Ultimately, we will prove that everything NASA has told us about the moon landings is untrue.

In fact, most of the pictures offered by NASA were created from the moon landing footage filmed here on earth. In the following chapters, you'll see mountain background scenes. These scenes are mostly from astronaut training sessions in New Mexico and Nevada. This explains why NASA does not provide many pictures related to the moon landings. They are afraid someone will match the scenery and figure out the Moon landings were faked. NASA has other reasons to hide most of the Moon landing pictures. Every time someone proves something is wrong with one of their photos, they are continuously forced to update the images with corrected versions. This is time-consuming and very expensive since all the American history books and government reference manuals need to also be changed with the new edited pictures.

For example, some of the original films and photos showed the Moon Rover with air tires. It has been proven that those air tires would have deflated in a space vacuum under the pressure. Today, the photos displayed by NASA show the Moon Rover with wire band tires. The air tire photos have been removed. To cover up the air tire problem, the original air tires, from the Moon Rover on display at the Kennedy Space Center, have been replaced with wire tires as well .

Award winning British photographer David Percy is convinced that the pictures are fakes. He says the shadows could only have been created with multiple light sources and, in particular, powerful spotlights. also believing some of the mistakes were deliberate. Because "whistle-blowers" were trying to reveal the truth, they purposely left errors.

If NASA was telling the truth about their alleged Apollo Moon Landings, they should be able to demonstrate that their photos are not doctored, full of trickery, or complete fakes. Nonetheless, NASA and the Apollo Astronauts are reluctant to talk about them. Their refusal to answer these questions, and provide any practical explanation about the photos, is just another strong sign that the Moon landings were a hoax. After completed this extensive research, it was obvious why NASA has been refusing to respond to questions about the alleged manned missions. Responding would only incriminate themselves to the deceitful activities.

For example, how does NASA respond to this question: If Neil Armstrong was the first man on the moon, then who shot the video of him descending the ladder and taking his first step onto the moon surface? The original specifications of the external cameras mounted on the lunar module could not have taken this picture. Besides, to take a snapshot of himself, Armstrong would have had to take his camera off, throw it out the door, and perfectly adjust it with some special remote feature, which this camera did not have.

The Apollo Spacecraft Landings Were All Filmed At the Same Location

NASA's own evidence again confirms that the moon landings were a hoax. The Apollo Moon landing films show the landing site scenery was the same for the astronaut's descent on to the moon surface from the LM spacecraft. Compensate the camera angle, and they are identical matches. Everything, including the surface landscape, background scenery, and distance correspond. This discovery also confirms the filming must have been taken at the same location. This was most likely the Kennedy Space Center because it matched the indoor training simulators there. These findings can be validated by simply reviewing each of the Apollo mission films where the astronauts are seen descending onto the moon surface for the first time. Here is a list of a few NASA Apollo Mission Films: the Apollo 11 ID# a11.v1094242, Apollo 14 ID# a14edjump, and the Apollo 15 ID# a15.jimststep. Samples of these NASA film clips are shown below for comparison.

These films have been around since the 1970's and visibly indicate that the Moon landings were faked. Why hasn't the authenticity of these NASA films been questioned before? We know Americans were afraid to speak up out of fear for their lives, but what about other countries? One would have to assume that other countries felt it was too difficult to prove NASA's claims one-way or the other. They must have believed it was impossible for their scientists to translate the data because nothing made any sense.

Apollo 11 firt step on moon NASA film #a11-v1094242

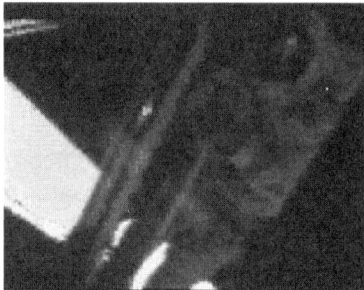

Apollo14 First Steps NASA ID# a14edjump

NASA's Custom Designed Camera Helped Prove the Moon Landings Were Fake.

Thanks again to NASA, they have provided everything needed to prove the moon landings were a hoax. NASA records show they made a small adjustment to all the cameras used on the Apollo missions.

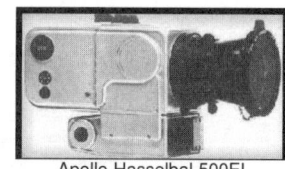
Apollo Hasselbal 500EL Data Camera

NASA claimed to have etched all the photo lenses of the cameras taken to the moon with a little "+" sign. This sign is referred to as a crosshair or reseau-line. Therefore, all Apollo video footage and pictures had to display a Crosshair pattern as shown in the picture to the right. Also, according to NASA, the number of Crosshair patterns and sizes must remain the same on every image.

Camera Crosshair Pattern

To calculate the distance, NASA claimed that the lines were etched into the camera lens. Incidentally, that is impossible. To take a special stereo picture that could determine distances, you would need two cameras set up! NASA's real goal was to achieve just the opposite affect. To actually distort the distance of backdrops and miniature objects in the photos, the crosshairs ("+") were added. To fake the moon landings, the photographers needed to distort distances, so the boundary limits of the various Apollo training simulators could be used.

After understanding the basic alteration techniques used by NASA, the role of the crosshair, and all the so-called "Moon Anomalies", can easily be explained as simple photographic trickery.

The Crosshairs Prove NASA's Photos Are Fakes

Since the crosshairs were sketched into the lens of the entire camera, there should be only one distinct pattern in all photos taken by the astronauts on the moon. This picture is an example of what crosshair patterns should look like in all photos taken on the moon.

In order to prove that NASA photography was doctored and to create the illusion of being taken on the moon, understanding the logic behind the crosshair patterns is critical. To better explain, the picture to the right was shot with cameras the Apollo astronauts supposedly carried with them. The image is a vivid example of the crosshair patterns that were sketched into the camera lenses.

NASA ID# as15-85-11507

Evidence NASA Forged Moon Landing Photos was Discovered

Examples of irregular Crosshair patterns would indicate that NASA was doctoring-up the Moon landing photos. There are thousands of inconsistencies where the wrong number and size of crosshairs are visible. Some Apollo Moon landing photography completely lack crosshairs. Many of these crosshair inconsistencies are the result of NASA doctoring the photos to cover up lighting hot spots, levitation harnesses and the miniature sized stage props. The photo below is a perfect illustration of these irregular crosshair patterns and the remainder of this chapter will detail how we can prove these photos are all fakes.

NASA ID# a15station6med
Very iIrregular crosshair pattern-indicating forge photo

Missing Crosshairs

In the next set of original NASA photos, the crosshairs are missing completely from the pictures. According to NASA's own camera specifications, this is not possible unless someone tampered with the pictures and removed the crosshairs because the crosshairs were etched into the camera lenses.

NASA ID#10075864

The notion that someone must have tampered with these photos makes absolutely no sense at all. If these photos are missing crosshairs and indeed tampered with, then it had to be someone working for NASA because they released the original photos. Because, at this time, digital formatting was not invented, it would have taken a whole team of photographers and artists to remove the crosshairs from these pictures. Then this team of photographers would have polished the picture to look like an original. Why would NASA even do such a thing? They wouldn't.

These original NASA Moon landing photos provided a valuable clue, which revealed how NASA photographers produced the fake moon landing photography. There is a explanation as to why many of the alleged moon landing photos were missing camera crosshairs. It's the result of a very common form of camera trickery used back in the 1960's and 1970's by the movie industry. NASA was using a 3-step process to produce Moon landing photography

3-step process to produce Moon landing photography

NASA used a simple 3-step process to produce the fake Moon landing film footage. First, they filmed astronauts on Earth pretending to be on the moon. Secondly, they edited the film. Third, they filmed the edited movie with a different camera. It was necessary to use the 3-step process because of the amount of editing that would be required on the original film footage. Because, while filming, there would have been too much commotion. It would have been common, while the astronauts were being filmed, for stage crewmembers, or the astronaut's family members to accidentally step into the view of the camera. During the outdoor filming a bird passing by could also be caught on film.

NASA # ap17-KSC-72PC-379

There are several significant clues that solidifies the theory of the 3-step process for photography trickery. First, there are two distinct types of Moon landing photography. The pictures that contain crosshairs and ones that do not contain crosshairs. This indicates NASA used one camera for the original movie film footage without crosshairs, and then took another picture of that original film footage with another camera containing crosshairs. The original movie camera was usually taken in color, this is evident of the large number of colored moon photos that do not contain crosshairs.

If a movie camera with crosshairs was used for the original film footage, it would have made editing extremely difficult and in many cases impossible. The final film footage in most cases was taken by a black and white cameras, with a light filter. This explains why there is an abundance of black and white mirror images of the colored moon landing photography found in NASA's archives.

Below are two different original Moon landing photos obtained from NASA archives. These graphics confirm the 3-step process of photography trickery. The picture on the left is the original colored film footage with no crosshairs present. The picture on the right is a black and white "mirror image" of the original film footage now containing crosshairs.

Crosshairs Were Added Later to Original Film Footage.

The two pictures below NASA claims were both taken on the moon at the same time. However that would not have been possible since a different camera took each picture. The picture on the left was taken with a camera the did not have crosshairs etched into the lens. The picture on the right was taken by camera with the crosshair pattern etched into the camera lens.

The photo on the left is the original film footage taken without any crosshairs. The picture on the right is the film taken of the original photo by a second camera with crosshairs. This explains why the second picture is an identical mirror image of the first photo, with crosshairs added. There is no magic as to why many Moon landing pictures are missing crosshairs; they never had any to begin with.

Crosshairs That Have Objects In Front Of Them

Here are more evidence proving that NASA doctored the Moon landing photos. Many objects mysteriously appeared in front of the crosshairs. These next images illustrate this condition. In these pictures, the crosshairs are hidden behind focused objects. The obvious conclusion is that NASA added certain items, over the crosshairs, into the pictures. Once the items were added, NASA artists airbrushed over the film.

There are many examples of crosshairs being behind objects. For example, items like the radio equipment, the Rover Vehicle, and even the astronauts themselves. In these next Moon landing photos, if the crosshairs were etched into the camera lens, how could these possibly cover the crosshairs?

NASA ID# as16-107-17

NASA ID# as16-107-17

GPN-2000-001149

This can only mean that the astronaut and other items were pasted into the scenery pictures. Notice how the crosshairs did not disappear behind any brightly lit objects. These objects are clear and dark. If there was intense light reflecting off white objects, by suggesting bleeding around the crosshair, one could try to cover this up because of saturated film. Thus, the crosshair would be detached. This is clearly not the case here, and these pictures can only consider evidence of the Apollo Moon landing being faked.

More Examples Of Crosshairs Behind Objects:

In this next picture, parts of the crosshairs have disappeared. This too would be impossible unless the film was tampered with in some other way. The crosshairs should be visible and not hidden behind objects in the pictures.

NASA ID# as16-107-17

NASA Photo ID# a15.1433821

Extra Crosshairs Also Prove Moon Landings Were Faked

More proof that NASA faked the Moon landings is loads of illustrations contained extra crosshairs. Under close examination, these next photos are also composite pictures, which is a form of photographic trickery.

This is where NASA would take backgrounds and lunar surfaces from other scenes. Once they had the backgrounds and surfaces, they cut them into small pieces, and rearranged them to create completely new scenery. For instance, NASA would fill in craters, so that they came across as a totally different place on the moon. This next photo illustrate these technique NASA was using.

NASA ID # a15dmh11284-8
Proof NASA was cutting and pasting pictures together

At first, the above photo appeared to be fine. However, once examined under a microscope, the pictures that NASA claimed were taken on the moon contain many miniature crosshair. Because of the large number of misplaced crosshair, this could only be a direct result of NASA photographers composing this Moon landing scene from many pieces of other images. Below is another example to confirm the above moon landing photo was not some type of defective photo damaged during development.

NASA ID# a17.1682021

More Samples Of Composite Pictures Containing Many Extra Crosshairs.

NASA ID# a15station6med excessive crosshairs

NASA ID# ap151463033 Crater half filled-in

Same Crater Used In Manufacturing A Totally Different Moon Landing site

**Excessive crosshairs where crater is filled in with other photo pieces
NASA ID# a16pan1232508**

CHAPTER 10

Proof Stage Lights Were Used To Produce the Moon Landing Photography

This is a display photo used only for illustration purposes

Chapter 10
Proof Stage Lights Were Used To Produce the Moon Landing Photography

In the pictures, supposedly taken by astronauts, most people would agree that the shadows are clearly not as one would expect from pictures taken on the moon. Using special lighting equipment and retouching techniques, NASA constructed studio quality images.

For example, in the middle of dark shadows, the lighting of certain objects can be seen clearly lit up with bright lights. The objects in the shadows should be pitch-black since there is no air to scatter the light on to the moon. In many pictures, in the shadow of the spacecraft and moon rocks, several astronauts were shown brightly illuminated. Even when in total darkness, the American Flag and the words "United States" always seem to be brightly lit.

In this chapter, the following pictures will confirm that the astronauts could not have taken these pictures without additional lighting. Even though, according to NASA, the astronauts only source of lighting was the sun. Supposedly, the astronauts were sent back and forth to the pitch-black surface of the moon six times and never thought about bringing a flashlight. This makes no sense. It's simply ridiculous.

Photo from inside the KSC Florida display

Vividly, the astronaut's pictures were of the highest quality and were considered some of the best photography known to humankind. The selected target and the required angle of lighting could only mean one thing; extra light sources were used with pinpoint precision. Even a very powerful glare from the moon surface could not produce lighting with such accuracy.

Mei Yang, (age 9), one of the younger Internet Whiz kids from Japan has a theory on the shadow and lighting discrepancies. Mei's goal was to be one of the first humans to actually step foot on the moon. On January 14, 2004, President George Bush Jr. announced to send men to the moon. Mei feels that by the time she is old enough to become an astronaut, the technology may be available to bring her to the moon. Mei's theory is, "The reason the important items in the pictures seem highlighted by stage lighting is because they were." This theory goes to show you that the most obvious answer is usually the right one.

In most of the moon landing pictures, multiple light sources were indicated by various shadow heights and directions of travelling shadows. Many shadows were not cast in the same parallel direction. Under close examination, most of the mismatched shadows show no uneven surfaces, or irregular ground slants. Additionally, they have enough height to be seen as parallel.

The picture on the right is a perfect example. The astronaut is lit from the back, yet the flag is lit from the front. If the flags were transparent, the cross-member rod would be visible through the center of the flag as a solid black line.

20135257

There is simply no way NASA's photographs could have been taken on the moon. The images are too perfect. It would take experienced photographers hours just to get the photo that flawless. Nevertheless, NASA wants us to believe that the astronauts took these perfect pictures without viewfinders.

NASA has repeatedly tried to dismiss the strange lights in many of the photos from the moon. Further, they describe the lights as unexplainable space anomalies. The set of pictures below provide us with an explanation to the mysterious space light anomalies. Here we can see how the light on the moon photos matches the light sources found in the studio pictures where astronauts (actors) were seen staging the moon landings. The training pictures were taken at the Kennedy Space Center, and the pictures on the right were supposedly taken on the moon. Yet when cross refferencing the different mission images, the light spots are identical. This indicates that all of these pictures were taken at the Kennedy Space Center. Yet in many photos, the light spots are identical

Example 1:

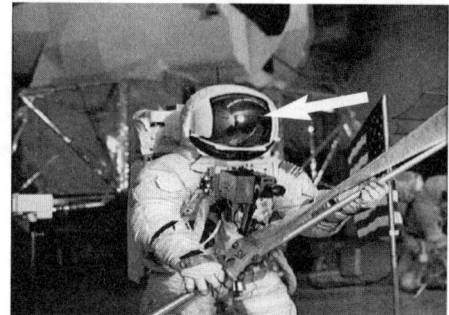

Example 2:

There are hundreds of Apollo mission snapshots with mysterious lights in the background. The illuminations were caused by the high-powered stage lighting. Some of the lights are so large that many people believe they are from the sun. Of course that's not possible. When taking a picture from the moon surface, the sun would appear to be gigantic.

So why would NASA publish these pictures? Their scientists also know these photos are not supported by modern day scientific principles. Is NASA again trying to leave additional clues of the moon landing hoax? Perhaps NASA believes that someday someone will figure it out. In general, since these are the original pictures taken by filmmakers, the reason they were published was by accident. One can just imagine how many years have gone by and no one at NASA is left. Furthermore, nobody can tell the difference between the original portraits and the doctored ones, so they release both!

Light Source Directions Help Match Training Pictures To Fake Moon Landing Photos

This next picture tells the whole story. Not only does the stage lights match, but the light reflections in the astronaut's helmet also are an identical match. Notice how the astronauts are in the same position, and the landscaping is the same on the moon as it is in the training picture.

Earth Training Missions

NASA's Moon Landing Picture

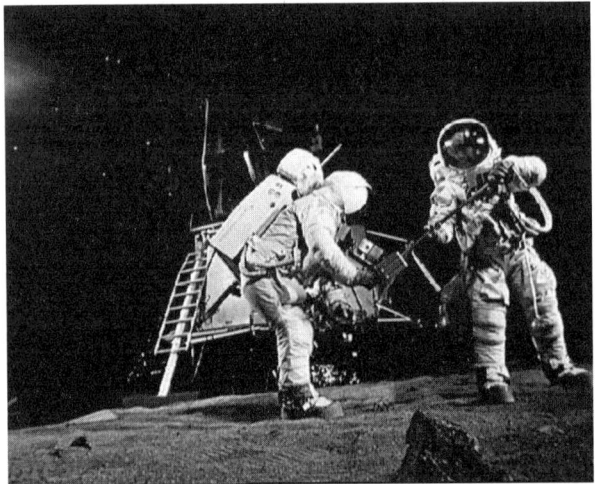

10075214

NASA was simply using a dual function camera with a day-to-night exposure filter. In the picture to the right, the ships landing pad has turned from red to black, and the astronaut's face is no longer visible. To add to this, the astronaut's space suits no longer look like they have moon dust on them.

Most importantly, the stage light, once visible in the upper right hand corner, has completely disappeared. Those engineers were something else back then. This stuff was very good for its time.

Astronauts Boot Have a 2-inch False Bottom

Below is another example of how the light source directions seen in the astronaut's moon landing pictures match the lights at the astronaut's training locations here on earth. Once this astronaut steps over the leg of the equipment and turns around to face the camera crew, those bright lights against the back wall are going to reflect off his helmet visor. Just as they are seen in the matching moon landing photo.

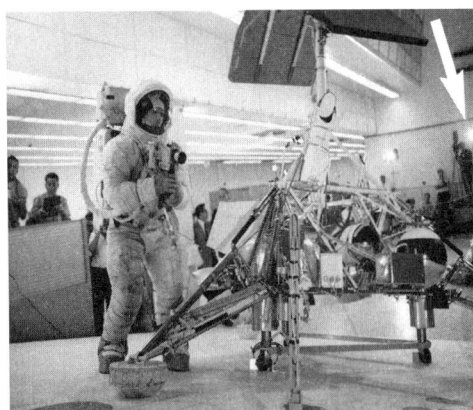

ap12-KSC-69PC-546

"Surveyor 3" NASA ID# 10075419

Also as mentioned earlier, we can see in this training photo, the Surveyor's legs are propped up. Also the astronaut is wearing Moon Boots with a 2-inch extension, to levitate him. The boot extension can be seen at the half way point of the gray area at the bottom of the boot. The intentional levitation of the astronauts and the Surveyor 3 equipment was to allow room for NASA's photographers to pass the objects into a different picture with the moon dust added to the identical hill scenery.

Many Apollo photographs show lighting diffractions (hot spots) as well as darkenings toward the horizon. In an airless environment with no spherical water droplets, sunlight would not have produced these "hot spots" on the moon. Even if there was a meteorite impact with tiny glass particles from the lunar soil, a condition known as a Heiligenschein, there still wouldn't be "hot spots". NASA claimed that the astronauts brought no other source of lighting with them. If there was no light source, what caused all these "hot spots" in the photos?

NASA ID# 10075611

There is an answer for so many "Hot spots" seen in the alleged moon landing pictures. Remember, earlier we discovered evidence that most of the Apollo moon landing filming was done at NASA space centers. This discovery explains the reason for some of the lighting "hot spots". The "hot spots" were caused by both the stage lights themselves, and the reflection of the powerful stage lights off the backdrop wall of the movie stage. This is exactly what would be expected. It's that simple!

To help explain this theory, one of the WhizKids, Muhammad Saudi, age 11, from India, developed a simple at home experiment. Her experiment helps explain the "hot spot" problem NASA faced with their faked pictures.

Here Is Muhammad's Experiment or Whiz Kid Experiment #6.

At nighttime, Muhammad said to place a lamp in the middle of a room. Next, turn off all other lights. You should see a bright reflection of the light bulb on the windows. This is known as a "hot spot" (or simply a reflection). The reflection on the window shows the problem NASA was having with their stage backdrop paintings and the black curtains they used when filming the moon landing pictures. This condition can be seen in many of the Apollo mission portraits.

20128963

To remove the "hot spots", NASA tried several methods. They attempted to airbrush over the light spots. NASA also tried to move the stage lighting to a different location and then take another picture of the same scene. After that, they simply cut the dark section out of the second picture and pasted it over the lighting "hot spot" in the first photo.

Proof That The Astronauts Carried Lights With Them

Despite NASA's Claims, More Evidence That The Astronauts Carried Lights For the Photography

NASA ID# 20147860

NASA ID# 20147908

Overhead Stage Light

There are hundreds of NASA photos that display evidence of stage lights being used on the moon surface. For instance, these next series of photographs show a rapid increase in light intensity. From this, the astronaut is lit from above. Was this astronaut in the wrong place at the wrong time? Was he zapped by a solar flare?
On the moon surface, a solar flare would have cooked this astronaut in his space suit. When studying these pictures in combination with other moon landing photos, the true nature of the strange light source is revealed.

Overhead Stage Light Increases

ap12-49-7278   ap12-49-7281   ap12-49-7319

Light on Moon match those in NASA movie studios:

To illustrate, in this next shot, the astronaut's helmet reflection contains a six socket lighting assembly. From above, this six socket lighting assembly would explain the strange lighting condition in many of the moon landing photos.

NASA ID # ap12-49-7281   **Six Panel Light Bracket**   NASA # as11-s69-31235

This explains why NASA never put headlights on the Moon Rover. The headlights would have constantly been reflecting off the movie stage backdrop paintings and walls. This would have triggered "hot spots." After going back six times, are we to believe that NASA engineers never thought of attaching a headlight to a vehicle that traveled along a rough pitch dark terrain like the moon? This also explains why the astronauts were not allowed to use flashlights on the moon stage. They too would have reflected off the background walls of the stage, which would have created more "hot spots" on the film.

Proof Stage Lights Were Used To Produce the Moon Landing Photography

Stage Light Can Be Seen Mounted Up Above

These next series of illustrations also reveal stage lights were used to produce moon landing pictures.

NASA ID# 20147789

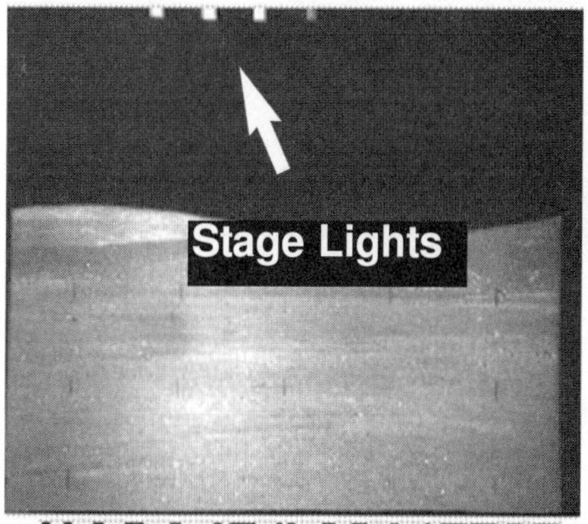

NASA ID# 20147787 — Stage Lights

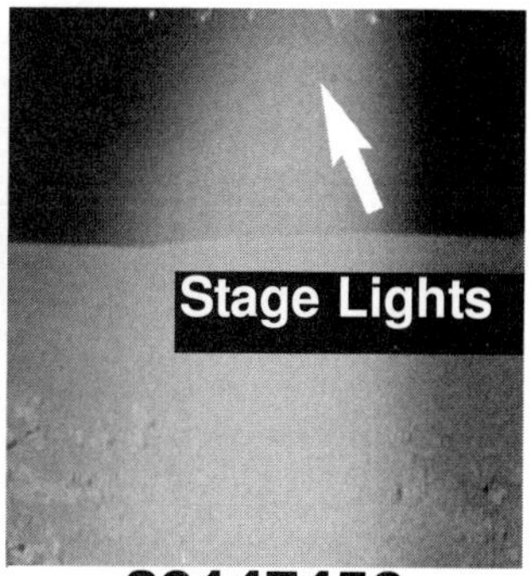

20147456 — Stage Lights

Six Panel Light Socket
20147456

Close-up Of Stage Lights

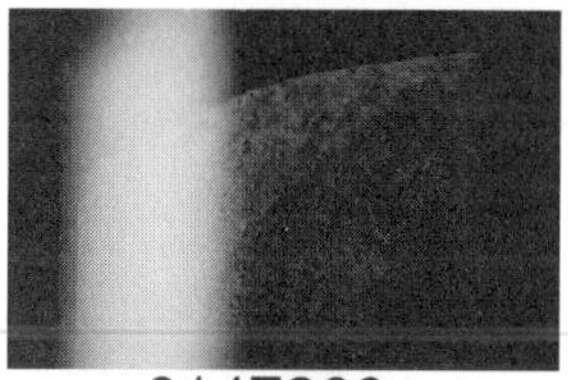

20134616   0147260   20147315   20147456

STAGE LIGHTS Hidden in Moon Rocks   The next picture is the most revealing of all. One of the stage lights can be seen hidden inside one of the moon rocks.

NASA ID # 20122176

Light Socket Found In Moon Rock

10-7

The Effects Of Stage Lights On Moon Landing Pictures.

There are many pictures that exemplify the use of stage lights. Until now, most people mistakenly thought the lights above and behind the astronauts were the sun casting a halo. These are often seen on earth, but not on the moon. Furthermore, halo effects can only be viewed through atmospheres and the moon does not have one. For this reason, the moon landing pictures are phony. To help illustrate this point, below are a few visual renderings.

20121864   20121863   20117351   20122439   20147895

20122441   20122527   20147952   20133850   20117363

Since the sun was the only source of light on the moon, the evidence of multiple light sources, as seen in these pictures, also proves that they were made on a movie set, which used many light sources.

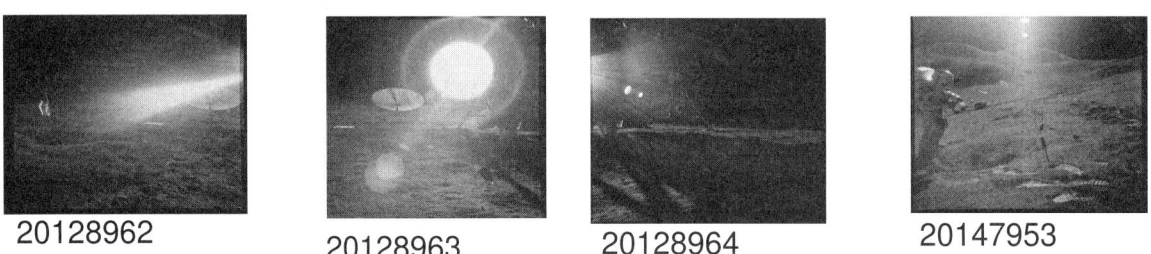

20128962   20128963   20128964   20147953

Evidence Of NASA Removing These Stage Lighting "Hot Spots".

The representations below are from NASA's archives. When you put them together, in a group, they offer a perfect example of how NASA photographers used the cut and paste techniques to remove "hot spots".

NASA ID# as11-5863-69

NASA ID# as11-5863-5

NASA ID# as11-5864-69

NASA ID# as11-5863-68

Once NASA photographers scam was discovered, it was very easy to see this type of cut and paste photography in many of the alleged moon landing pictures. In many cases, all the photographers had to do was darken the background as demonstrated in the two NASA snapshots below.

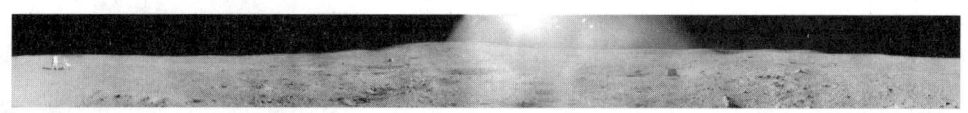
NASA ID #: a14.jpg Sample Of How Editors Darkened the Backgrounds

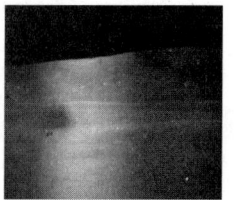
20147789

Before this book was published, on NASA's website, they provided samples of their doctored "hot spot" photos . Unfortunately, by now, it is no longer available to the public other than in this book.

Diverging Shadows

This next figure of Neil Armstrong placing the US Flag on the moon makes you proud to be an American. However, once a person stares at the photo, they realize it was filmed in a studio on earth.

This picture has multiple light sources. This is obvious by the shadows traveling in opposite directions, which is marked by the arrows. Since NASA claimed the Astronauts brought no lighting equipment with them to the moon, and the sun was the only other source of light, this photo is bogus.

NASA ID#20130716

Below are a few more photographs showing shadows pointing in different directions allegedly from the moon's surface. Under close examination, none of the mismatched shadows were caste on extremely uneven surfaces, which in some rare cases could result in a light divergence.

 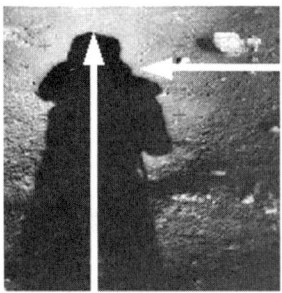 

If the light for the scene was not from the Sun, the conditions in these moon landing photos would be expected, but the light is coming from artificial light sources. To take these images, one would need a special focused light source producing a beam that spreads out from its rather close source and produces nonparallel shadows. A perfect example of such a light source would be stage lights found on a movie studio.

Multiple Astronaut Shadows Also Confirm Stage Lights Were Used For Moon Landing Photography.

Another valuable clue that multiple stage lights were used is the discovery of a double shadow in the moon landing photography.

Notice how in the picture to the right, the astronauts have two shadows rather than one. For an object to cast two identical shadows as seen in this picture, two powerful light sources would be needed. If we are to believe the sun was behind the astronaut in this picture, then the moon's reflection would have been traveling in the opposite direction, and the two would have canceled each others shadow. With only one light source direction, it proves this picture was not taken on the moon, but here on earth in a movie studio.

NASA ID #a12vr_1162229.mov
Double Shadow

10-10

Shadow Cut and Pasted into Moon Landing Photo's

This NASA original Apollo 15 photo is undoubtedly fabricated. The crater is filled in with irregular crosshairs and the shadows are going in multiple directions. There is another very big clue that this Alleged Moon landing photo is a fake. The Astronauts shadow is not his own. Although the shadow may look very close, keep in mind that it must be an exact image of the astronauts' position. This astronaut's shadow was added to this picture from another scene. What was NASA thinking?

Astronauts Shadow does not match astronauts position

Example of composite photo obtained from NASA. Photo ID # a15station6med

Shadow Traveling Three Different Directions

The following NASA wide angle picture also helps show several light sources were being used to produce the moon landing photography. If this scene was lit by natural sunlight, the shadows would be falling in the same direction. However, in this shot, they go in three different directions. If this scene were lit by natural sunlight, or a reflection from the moon surface, the shadows would still be traveling in the same direction since the light would be naturally combined. For this reason, multiple light sources must have been used.

The figure below sums up the evidence of several light sources being used. Otherwise, how would you explain the portion on the right side of the picture below as being described as a 'spotlight'?

Multiple Shadow Directions

NASA ID# AS12-47-6987

201286902

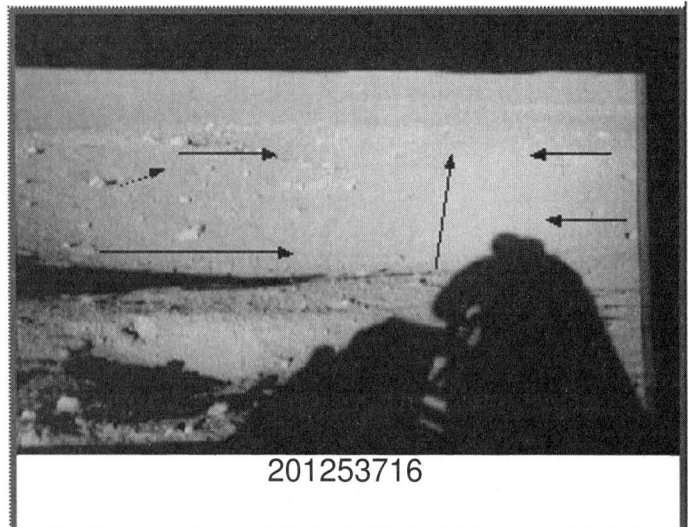

201253716

More Shadow Inconsistencies

Here, before the astronauts head out for a hard days work on the moon, the Apollo 14 Astronauts get in a quick game of golf. In the different visual aids, notice how objects were removed and sizes were adjusted.

The graphic on the right looks fine except for the fact that the astronauts were playing golf on the moon. The middle picture has changed a little. The astronauts left, and a second, smaller flag, was added. Even though the astronauts supposedly had left, they are still there. The figure on the right shows the larger flag has been removed. This is a perfect example of a composite photo used to fake the scenery in a photo.

This event simply could not have taken place on the moon, yet NASA broadcasted live to millions of eyewitness viewers on TV. The logical conclusion is that NASA must have broadcasted a prerecorded moon landing simulation on live TV, which was easy to do.

Most likely, at this point, NASA was tired of the hoax, so they tried to reveal to the public that the moon landings were faked. Unfortunately, NASA did not understand the technical revolution was just starting to get underway and most of earth's population still had an industrial way of thinking. For NASA, something even as obvious as this simply went undetected, and the Apollo hoax only continued to intensify more and more out of control. Oh well, NASA is ingenious. They will figure out someway to reveal the truth of their hoax. Who knows maybe someday they will secretly hire someone to write a book just like this one.

These images are the result of NASA combining various items with some very careful retouching. That miniature flag was a major key to deciphering how NASA managed to fake the entire moon landing. To distort distances of the background scenes (paintings), evidence suggested that miniature landing site props were used in the photos. These miniature props were necessary because the moon surface platform had less than 100 feet between each background wall.

Incorrect Lighting

In the picture below, notice how the shadow falls to the left. This creates blackness except for the astronaut and satellite dish. Clearly, two powerful light sources were lighting up the astronaut and satellite dish from the opposite direction. While it should be completely black, the astronaut is in full light. Additionally, the two objects are not projecting any shadows.

Light beams on back of astronauts head, yet shadow casting that direction. Proving multiple lighting source were used.

In These next two photographs, the astronaut is lit up on the side that should have a dark shadow. In the picture on the left, the astronaut is in a large shadow cast by the spacecraft, yet his entire body is still visible. The reason he is not surrounded in darkness is because there is more than one light source being used, which means these photos were taken on a movie studio.

More Proof NASA Doctored Photos

These next two images show a method NASA used to manipulate pictures to create the illusion of being on the moon. First, the photographers filmed the scenery. Secondly, when the picture was just the way they wanted, they took a picture of that scenery. This is noticeable since the picture on the left is missing the crosshairs. Evidence showed that this was because, for the original scene, the movie filmmakers did not have crosshairs embedded in the camera.

The picture on the right is a negative of the first picture with the crosshair included. Depending on the scenery, NASA occasionally took pictures of the original film to add the crosshair patterns that were supposed to be etched into the camera's lens. Another obvious clue these pictures were taken on earth is the visible lights. Without an atmosphere, there is no sunrise on the moon. Because of this, the light had to be caused by a stage light from the background of the studio.

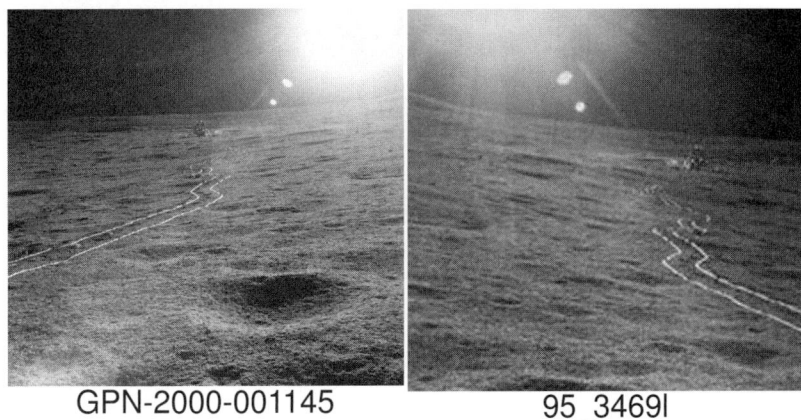

GPN-2000-001145   95_34691

As seen by the figure below, the lights are coming from behind the "black net like curtain" at the Lunar (moon) simulator studio outside the Langley Research Center. To build stage sets and have actors in space suits act out the moon landings, NASA went through much time and trouble. This will be covered later in the book.

If you are interested in seeing more cutting and pasting photos like these, go to NASA's web site at www.nasa.gov. At this site, you can find many doctored images. You will also find an Apollo image search feature for all the Apollo moon landing missions.

No Shadow

This Apollo 11 picture below indicates the extent to which NASA altered photos to create the illusion of being on the moon. The shadows are not parallel with each other, and there is no shadow at all from the flag. Under close examination, the astronaut's feet are cut in half because the astronaut was pasted into this scene.

Flag has no shadow
NASA ID# as11-40-5874

Other People's Shadows Found On the Moon.

Kids wonder away from parents and get onto the artificial moon surface stage at the Kennedy Space Center. The shadows in this NASA picture are of a young girl holding a purse and her little brother. They are definitely not astronauts.

Girl With Purse
NASA ID# 20147711
Boy

ap17-KSC-72PC-379

NASA ID# 10074013

CHAPTER 11
Astronaut Training Photos Used As Moon Missions

Chapter 11
NASA Used Astronaut Training Photos

Earth Astronaut Training Photos match those photos NASA claims were taken on the Moon

After gazing over thousands and thousands of moon landing photos, a remarkable discovery was made. There is a common connection between the Apollo Mission Training Simulator photos taken on Earth and the pictures the American Government claims to have taken on the moon back in the 1960's and 70's.

Overwhelming evidence suggests NASA's photographers simply used common methods of photo manipulation to convert the Apollo training photography into realistic looking images. After discovering the connection, with just a glance, most of the photos can easily be recognized as fake.

To create the dark and eerie appearance of the moon, a custom light filter lens was designed for the camera called a "Lunar Day Lens". This type of light filter lens is commonly called a Day-to-Night Filter and is used often in the movie industry. A Day-to-Night Filter and a piece of Smoke Plexiglas can produce a very similar effect on film footage. Have you ever noticed how a country western movie appears to have a bright yellow look to it? This is because the custom yellow lens has attached to the cameras during filming. This special yellow lens attachment is referred to as a "Day to Country Western Filter."

Camera Hasselblad El Data and Lunar Day Lens attachment

Smoke Plexiglas effect

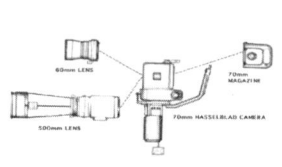
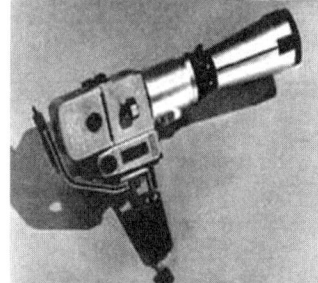
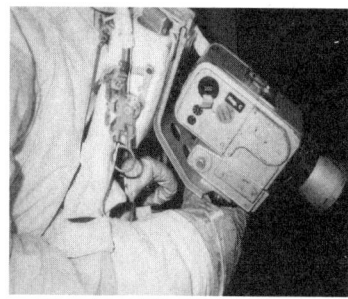

NASA used a Day-to-Night Lunar Lens filter to create a darkened image of the moon, while filming the astronaut conducting missions on Earth in the middle of the day. This explains why NASA used bright white astronauts space suits and most of the equipment. The white color made it easy for NASA's photographers to offset for the Gamma Light reduction. This would also give the astronaut's space suit the appearance of being covered in moon dust, when the light was filtered out. All NASA had to do was touch up a few spots in the black space background and the training mission photos would look as if it were taken on the Moon.

To illustrate this theory of how the camera light-filtering lens worked, the next set of slides have been put through the process. There you have it a picture of the astronauts working on the moon. Could anything be easier?

Before   After

NASA Cameracrew adjusting Day to Night filter
NASA ID# AS12/10075363

Evidence from NASA reveals "Day to Night Filters" were used.

For example, this set of NASA pictures shows how one of the Earth Training photos was filmed with the Cam-fov light filter. Soon after, the pictures were claimed to have been taken on the moon. During the first few Apollo Missions, this form of photography trickery was used many times.

Earth Training Photo's Identically Match The Moon Photography

Attached to the lens of the camera was a custom light filter to create the moonlight appearance. This light filter is commonly called a Day-to-Night Filter and is used often in the movie industry. A Day-to-Night Filter would produce an effect similar to placing a piece of Smoke Plexiglas in front of a common home video camera lens.

Apollo 11 Training at KSC

Apollo 11 Landing Site

Earth at KSC Training

Apollo 11 Landing Site

The Surveyor-3 'Hoax' Film Footage

The Surveyor-3 pictures at the Kennedy Space center reveals the film footage was faked. At first, it seems like the typical moon- landing hoax picture setup. First, the astronauts were filmed with the equipment. Next, the photographer filmed the equipment on an artificial moon soil surface (*A sand and charcoal ash mixture*). Lastly, the Astronauts were pasted into the completed moon surface pictures. This type of photography trickery is nothing new, it has been around since film was developed.

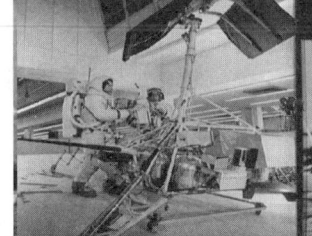

What's disturbing about these pictures is observing the astronauts taking a picture alongside the Surveyor-3 on earth. NASA's records show that during a unmanned mission Surveyor 3 had been sent to the moon. Additionally, this information exposed that Surveyor-3 arrived on the moon three and a half years before astronauts. When it got there, it was broken. Yet, we see the astronaut taking matching pictures here on earth and learning how to operate a broken piece of equipment. This was supposed to be on the moon. This photo and NASA's story make no sense. Why would the astronauts go through intensive training on a piece of equipment that no longer works? Besides, if the Surveyor III was on the moon, why would the astronauts be posing for a portrait that has the same sloped landing scenery that was only discovered on the moon several months later?

Either NASA faked the moon-landing picture or they had a time machine. It's safe to bet they forged the pictures on earth. In addition it is obvious the human race will not survive long enough to build a time machine. If they did, someone by now would have returned from the future and encouraged everyone to put a stop to the continuous global devastation, which is leading to the extinction of humankind. Why doesn't NASA just admit the moon landings were fake? NASA should allow everyone to move on to bigger and better scientific achievements; For example we could focus on designing stable super conductors to harness the power of magnetic flux lines. Perhaps, NASA could fund the development of safe methods for accelerating the half-life cycles of nuclear waste and hazardous chemicals. Ultimately, NASA could master particle physics.

Matching Mirror Images

These next set of NASA photos reveal the astronauts' training landscape is an identical match to the alleged Moon Lunarscape (scenery). The discovery of these images by themselves is enough evidence to prove the moon landings were faked. How could the Astronaut's training photo rock formations and the slanting moon landscaping match the landing sites where they supposedly worked on the moon. Take out a few of the last minute items inserted into the scenery by NASA's artists and these Earth and Moon photos are at the exact same location.

NASA's Mirror Image   **Apollo 11 Landing site**   **Apollo 11 Landing site**

NASA ID# 10075214   NASA ID# j.ud.y.42   NASA ID# 10075290

Since the astronauts could not have taken the training center to the moon, they must have brought the moon to the training center.

Here is how NASA produces these Moon landing images from the astronaut's training film footage. The film footage on the left was first taken during the astronauts training at the Kennedy Space Center. Then the photographers produced the image in the middle by filming the original film footage with a second camera, with a Day to Night filter attached, which explains why it is a darker mirror image of the original picture.

The picture to the far right is the same image except NASA has converted it to a Moon landing photo, by filtering out more light and having their artists touch up the photo.

Original Film Footage   **NASA's Mirror Image**   **Apollo 11 Landing site**

 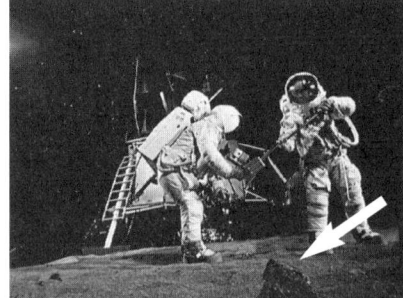 

NASA ID# a11.s69_32243   NASA ID# 10075214   NASA ID# j.ud.y.42

Producing the Live Broadcast Film Footage

To produce the Live Broadcasted Film Footage NASA's photographers simply took the light-filter and touched up film footage and blurred it with another filtering system.

After discovering how NASA produced the images, it is easy to see how they could have produced almost any Moon landing film the footage with a their colored cameras.

Live Broadcast

11-4

Straightforward Matching Photos

These next moon landing photos show how easy it was for NASA to use the Astronauts Training pictures to produce the moon hoax photography. Occasionally NASA needed to cut and paste a certain portions of their Moon Hoax photography.

There are many clues in the alleged Moon Landing photos that prove they were originally filmed at the Kennedy Space Center. First filming the astronauts mission activities in a moondust free environment.

Next the moon dust would be added and the real Lunar Lander Spacecraft and other moon stage props. Then NASA photographers would take photos with the astronaut's wife and daughter watching, and paste it into the moon landing scenery in the other room. Often this method of cutting and pasting for still pictures was used. It allowed astronauts to be filmed anywhere without having to wait for the scene to be set up with dust, from these pictures could be taken immediately.

There is strong evidence proving this Moon Landing photo was filmed inside the Kennedy Space Center Movie studio seen above.

ap17-KSC-72PC-379 -KSC

There are many clues that confirm NASA first filmed the Astronauts and then paste them into moon sceneries. For instance;

1. Notice how the astronaut's boots have a wide gray band at the bottom. Color is almost a perfect match to the moondust soil. The boots were intentionally designed this way to make the cut and paste process seamless. With a magnifying glass, the touch airbrush work can be seen in the alleged moon-landing picture. This explains why in many pictures it appears as if parts of the astronauts feet have been removed, on the uneven soil spots.

2. With just a quick glance, a crease in the moondust soil alongside the edge of the gray plate form can be seen. The crease was obviously caused by the moondust settled along the edge.

3. Fresh moondust, slightly dark in color, has been added over the Flag sandbox to cover up the raised edges. This darker moondust can be seen in many of the different moon landing mission photos. It helps create astronaut footprints and marks the walking path in the work zone. The outer area moon dust being a much lighter color and settled confirms this theory.

NASA ID# a17-20117331
Apollo 17 Moon Landing Photo

4. Even the flag is mounted straight upward, yet there are no astronaut footprints inside the sandbox.

5. Finally, you'll notice the two rocks inside the sandpit are an identical match to those in the Earth training picture and the alleged moon-landing image.

Below is a view of the NASA Moon Landing pictures where the astronauts feet have been erased during the photographers cutting and pasting process. Often the flag is mounted with no footprints in the Sandpit box or around the flag. In some cases, there is a footprint or two inside the sand pit on the moon. Further, these prints have sunk much further into the sand than the footprints surrounding the Sandpit box area. What was NASA thinking?

as16-113-18342

Evidence shows NASA photographers did a good job of disguising the cut and paste work on a rather smooth surface; however, they had all sorts of problems with trying to paste the astronauts into one of the moon-landing scenes with an uneven surface.

Movie Stage platform with SandPit Box for mounting American Flag

4- Dimensional Photography Deceptions

The objective of this 4-Dimensional experiment is to prove these next two pictures were taken at the same time, and by the same Astronaut. However, your first reaction may be, "it's impossible for these two photos to have been taken at the same time, and place, since one was supposed to be taken on the moon and the other on earth during the astronaut's training."

To prove this Moon landing photo was taken on earth during the alleged astronauts training, I have put together a visualization experiment. To the best of my knowledge, this is the only 4-dimensional thinking test of its kind ever put on paper. It is a lot of fun and I encourage you to share this experiment with your family and friends. They will be glad you did and they will probably thank you for opening their eyes.

Are you ready to take the 4-Dimensional thinking test ?

These two NASA pictures were taken at the same time and place

a14-s70-46153

First, see if you can find the flag in the picture on the left. Now imagine yourself being the astronaut standing in front of the spacecraft looking at the astronaut holding the American flag. You are getting ready to take a picture of the other astronaut standing near the flag in front of you. You are now positioning your camera and the movie producer yells "Action". Suddenly the lights go out and one of the large spotlights mounted to the LM spacecraft behind you is turned on. The spotlight is so powerful that it lights up the astronaut holding the American flag in front of you. It's now time for you to go to work. Your job is to take a really good snapshot. Remember, your photo shot has to be good enough to fool the world into believing it is real!

Naturally, as you're standing there you are a little nervous. Who wouldn't be; what if you get caught? Okay, your taking a deep breath, and are now trying to focus the camera but that stupid American flag keeps blowing around and getting in the way, this is making you even more nervous. Your hands are now beginning to shake, you say to yourself, I don't need this crap, and pull the trigger. Congratulations! You did it. It's a perfect match. See for yourself!

4-Dimensional Photography Continues.

Many earth-training locations can be matched to the alleged moon mission photos using 4-dimensional thinking. See if you can match this next set of NASA photos using the same techniques.

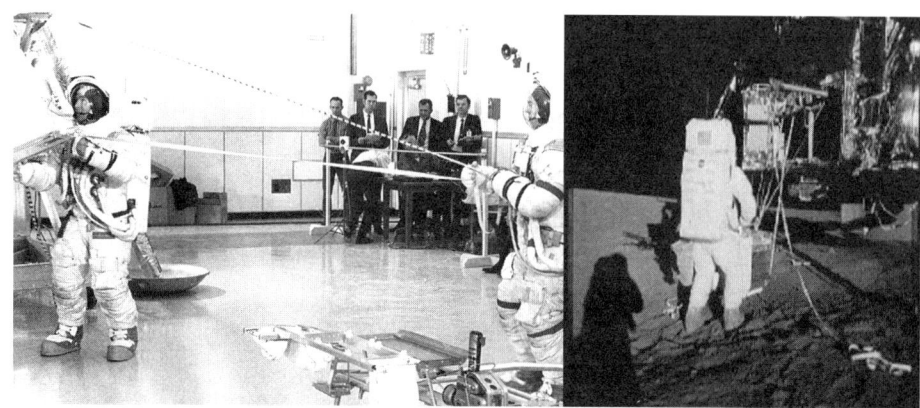

Matching Photos From Different Angles.

These next three images are special. It appears as if the armadillo animal seen in the earth picture on the left has somehow managed to make it to the moon and is now building a nest under the tool cart on the far right.

An armadillo on the moon, of course this would be impossible and makes no sense. It must be some simple mix up. Maybe the picture on the right that looks like it was taken on the moon is simply a result of a dark cloud

Apollo 14 Training
NASA ID# ap14-KSC-70P-498

From NASA Apollo 14 training
Photo ID# j.jpg

from Apollo 14 Mission
NASA ID# judy61

passing over the training area. It would then be very easy for someone in the NASA editing department to have mistaken the picture on the right as a moon mission photo. No that's not possible. Remember NASA claims only photos from the moon contain crosshairs and the matching moon picture on the right has crosshairs! Without a doubt, NASA's photographers simply pasted the moon background scenery to the matching earth picture to produce a moon landing hoax portrait.

CHAPTER 12

How NASA Created the Moon Scenery

Chapter 12
How NASA Created the Moon Scenery

This chapter contains some of the most compelling evidence yet. Furthermore from the evidence, a detailed analysis of the suspected Moon landing movie studios will be provided. It also includes proof showing how the Moon's surface was faked in the pictures, and the methods NASA used to create the illusion of the astronauts being on the moon.

Up to this point, almost all the moon landing photography contained some type of alteration, which eliminated the slightest possibility of human error, or an unexplainable Moon anomaly. To conclusively prove the Moon landings were faked, the remaining evidence in this book, will help bring all these pieces together.

This sounds like a tall order to fill, but considering the evidence, it's not going to be that difficult. In "today's world", if it were anyone other than the American Government trying to pass this photography off as authentic, people would never have taken them seriously. However, proving the American Moon landings were a hoax would never have been possible back in the 1970's, 1980's, and early 1990's. Back then, there was simply not enough material available on which to base an investigation. There were only a few photos in a limited number of science and fictional books. This explains why the American Government's Moon Landing hoax has gone undetected until now.

Unfortunately, for the Americans today, anyone within a few hours can search NASA's catalogs of photos on the Internet and see the fake Moon landing pictures themselves. It's no surprise that, when NASA made more Apollo Moon Landing photos available on the Internet, the number of people claiming the Moon landings was fake also increased. With just a glance, most of NASA's Moon landing photos can easily be recognized as having serious problems resulting from the photos being fabricated.

For instance, one very major problem with NASA's photos is that different landing site pictures have the same background mountain scenery. They were supposedly on different areas of the Moon. It appears NASA used the same scenery in many different landing site photos. Again, if it were anyone besides the American Government who was behind this, there would be no question that this photography is 100% fake.

For example, these next two images appear to have identical mountain backgrounds with totally different Moon landscapes. One is very rocky, and the other has a very smooth Moon surface. Typically, this would not be something one would be suspicious about. The astronauts could have moved further away from the mountain, and reached a very rocky landscape. Nonetheless, it was determined that this is not the case in these next two photos because the spacecraft has also disappeared.

 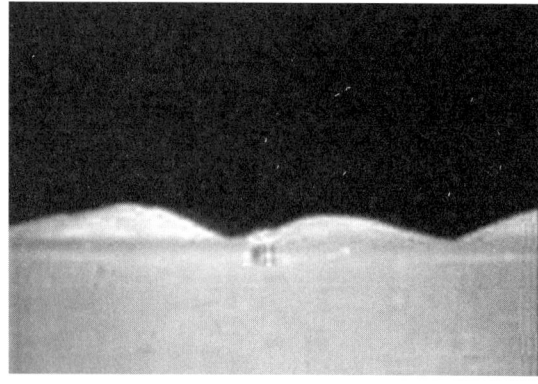

Since the astronaut's spacecraft was incapable of moving around on the Moon's surface after landing, then it must have been the mountains that moved from one location to another. The question now is: could it be that mountains on the Moon move around, or is this another clue as to what methods NASA's photographers used to produce the Moon landing film footage? The first step in solving this puzzle, would be to search for more evidence that exhibited this same type of identical backgrounds used in multiple Moon landing locations.

Identical Backgrounds With Disappearing Astronauts' Spaceships

These next sets of pictures have the same mountain backdrop, yet the Lunar Module Spacecraft is only visible in one of them. This seems impossible because according to NASA the LM spacecraft never moved from its original landing position. Also even after the mission, the LM base would have remained.

NASA Photo ID #
as16-113-18342

NASA Photo ID #
as16-114-18427

This is another example which supports the Sliding Scenery theory. When shooting separate pictures, NASA used the same artificial backdrop. The traditional photography "cut and paste" method can be seen here. The foreground and mountain scenery dramatically change color shades where the stage platform, and the background painting intersect. The scenery shows how artists used airbrushes to touch up the photos.

Disappearing LM Spacecraft

These next two Apollo 15 mission photos show NASA used the same mountain background for different locations on the moon. One image contains the LM spacecraft and in the other scene there is no ship.

Fake Mountain Background

NASA ID# AS15-82-11057

NASA ID# AS15-82-11082

Movie Stage Platform Area

Film Crew and Photographers Area

The mountain range on the left side of each photo is called the Apennines Front. The Mountain, Hadley Delta, is to the right. According to the design of the Lunar Module Spacecraft, the bottom section must have remained on the Moon's surface after the astronauts left. This means the picture on the right must have been taken before the ship had landed. On the ground, there would have been no welcoming party to take this picture before the astronauts' ship touched down.

LM Spacecraft Disappears Again

Below are two more pictures supposedly taken from the same spot in front of the Apennines Front Area, however they somehow manage to have different foregrounds. In the picture on the right the camera zooms OUT to get a wider view and the LM Ship disappears.

NASA ID# a15.1254512_dmh

a15pan1480225

Camera Zooms Out and Ship Disappears

If the camera had zoomed IN at the mountain then the ship would have disappeared. However the picture on the right is a zoomed out image of the mountain and the ship should have remained in the photo. Once again proving NASA was re-using the same moon background scenery for different landing sites on the moon.

Under close examination it was easy to determine that the airbrush touch-up was performed where the stage platform connects with the background scenery. As we go through the remaining pictures, see if you can find the intersection line between the Stage platform and the backdrop paintings. Hint: remember the stage platforms were mostly always less than 15 meters (50 feet) wide.

Disappearing Mountains

Evidently in NASA's alleged Moon landing photography, the Moon was a much stranger place than would be expected.

First, we have disappearing craters. Next, the Lunar Module Spacecraft magically disappears. Now, mountains vanish without a trace. It would appear NASA not only had the technology to land men on the Moon, they also had a secret cloaking technology that would make an item disappear. Below is an illustration of how hills disappear without a trace.

Of course, the most sensible conclusion would be that NASA's photographers used different sections of their artificial background mountain scenery to create the illusion of different landing sites in these illustrations below.

The first photo is a full view of the Moon background painting that was used for several different landing sites. NASA claimed the second image was a close up view of the "HILL E" section. Nothing seems out of the ordinary other than the foreground changed from rocky to flat.

The third picture was supposedly a different Moon mission. This mission was implemented at an entirely different location on the Moon. Besides the fact that the primary and secondary horizon lines are almost identical, the mountain to the right and left of "Hill E", has disappeared altogether.

System for Proving these are Identical Backgrounds

This next series of identical background pictures were supposedly taken during the Apollo 17 mission. However, on the right side of each picture there is a common curved light shadow pattern and a small dark crater built into the artificial mountain backdrop. All the pictures have the same background and also have identical landmarks.

Besides, if these pictures were really taken on the Moon, they would have a very frightening story to tell. Someone must have taken the pictures on the right with the spaceship gone, but how would they have gotten back to Earth. Perhaps one of the astronauts were left behind.

Apollo 17 Littrow valley

Apollo 17- Cochise Crater

Oh no look; an astronaut has been left behind! In the below picture on the left, you can almost imagine the shock the astronaut must have had when he noticed his spacecraft was gone. In the picture below on the right we can almost feel how much the astronaut is panicking as he dashes back to the landing site only to find no sign of his ship.

There is no need to worry; he will be okay and get back to earth. If the ship had left, we would still be able to see the lower portion of the ship, because it remains on the moon surface. The astronaut is obviously just a little lost. Who wouldn't be! It would be very confusing on the moon. As NASA has shown throughout the moon pictures, many areas look the same. No matter what side of the moon you're standing on.

Apollo 17 -10075973   Apollo 17 - 10075965

The story of the astronaut being left behind would only make sense if these photos were not taken on the moon. A more sensible theory would be that NASA used this same backdrop scene and then claimed that each picture was taken somewhere else on the Moon. The sudden straight line between the foreground movie stage platform and Background Mountains shows the same backdrop.

One way to determine if NASA used the same mountain scenery for several different Moon landing sites would be to perform an equal distance check. This could be done by simply measuring the distance between the two most distinguishable reference points, which appears to be present in all of the different landing site images. If the distances between the two reference points are equal in every picture, then it was the same scenery.

How NASA Created the Moon Scenery

Distance Check Proves NASA's Moon Landing Photos Are Fake

Below is the result of using a 1-inch equal distance test to determine if the different landing sites have the same mountain background? The distance testing confirmed NASA used the same mountain background scenery for different landing site locations.

Please note, there has been a third reference point added to confirm if the astronauts' Lunar Module Spaceship was moved in several of the Moon landing photos. Yes, the ship has also mysteriously been moved.

WHAT HAPPENED TO THE SHIP

Apollo 17 with Spacecraft

Spaceship disappear
NASA Photo # a17.1662658_dmh

NASA ID # 10075968 "kitty"

NASA ID# 10075974

NASA ID# 10075949

NASA ID # 10075968

12-7

Another clue proving the moon landings were faked is a startling difference between the Apollo 17 images taken from the LM window just before the Astronauts walked on the moon surface and this image taken just as they were leaving. When comparing them to each other, it's obvious that neither of these photos were taken on the moon. When the astronauts first arrived at the landing site there were many small craters spread out all over the place. However in the second image taken as the astronauts are leaving, most of the craters and a few very large rocks have magical disappeared, just like their ship did earlier.

Apollo 17 Landing site before stepping on to the moon surface.

Apollo 17 Landing site after Departure: Craters and Large Moon Rocks magically disappear.

NASA's two most common methods of faking the Moon Scenery

The Composite Moon Scenery

To make realistic looking Moon scenery, NASA photographers simply cut and paste pieces of various photos together. This enabled NASA photographers to create new backgrounds from other scenes. This composite form of trickery will be discussed later in this book in great detail. For now, here is a brief example of how NASA used composite snapshots to fake portions of the Moon landing films.

These next two photos, with identical backgrounds, hold many hidden secrets to NASA's photography deception. First, there were supposedly two different locations on the Moon's surface. These locations were Station 4 and Station 6. Furthermore, Station six's mountain scenery has a large crater in front of it. Now, look at the Station 4 photo. It has the same mountain scenery, but the crater has now been filled in. How can this be?

One important clue that proves these images were fakes can be seen in the second picture. The crater has been clearly filled in with small pieces from other pictures. This is confirmed by the large number of mis-matched crosshairs in the crater area.

The Sliding Background Moon Scenery

NASA also used a common method used also in movie production known as the Sliding Sceneries to create the illusion of distance on the Moon's surface. NASA alternated various large Moon scenery paintings behind an artificial movie stage platform, under the gigantic crane structure, constructed at the Langley Research Center. This system was ideal for filming the live action movies of the astronauts working at many different landing sites. Below is a photo taken back in 1968, of this very elaborate outdoor Moon landing movie studio. At the time of the alleged American Moon Landings, the facility was fully operational.

Langley Research Center moon Simulator-1968

How NASA used the Lunar Gravity Simulator at the Langley Research Center to fake the Moon landing Scenery

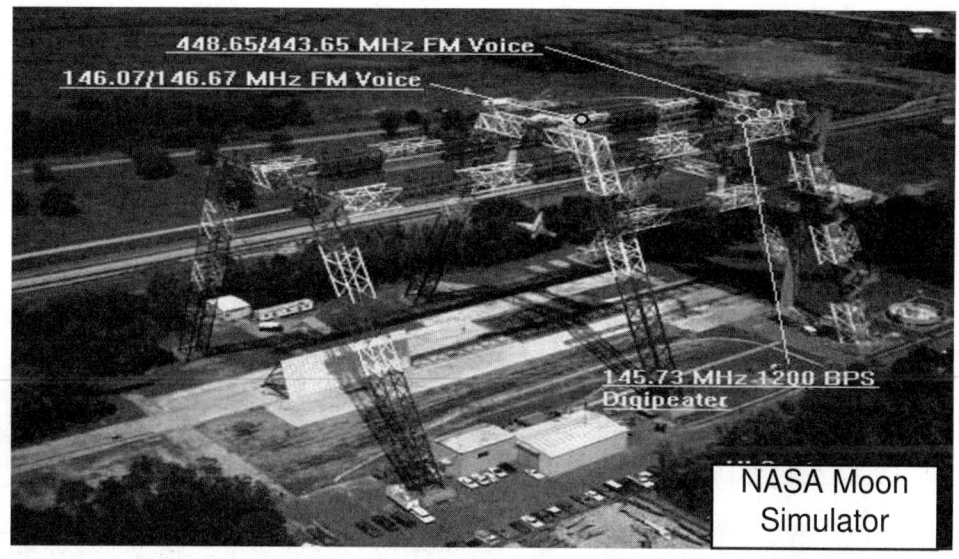

The Movie Studio At Langley Research Center

During the 1960's and 70's, NASA used gigantic crane structures, at the Langley Research center, to produce several portions of the astronauts' Moon Landing Missions that were later claimed to be real. At the time, NASA was disguising this elaborate movie studio as an Impact Dynamics Research Facility that was supposed to be testing the effect on aircraft crashes. However, the facility used a combination of pulleys, levitation cables, and sliding scenery for the background of the Moon landing pictures. NASA produced the most realistic looking astronaut Moon surface film footage ever imaginable. Under this strange looking gigantic crane structure, was where the most expensive movie of all time was filmed.

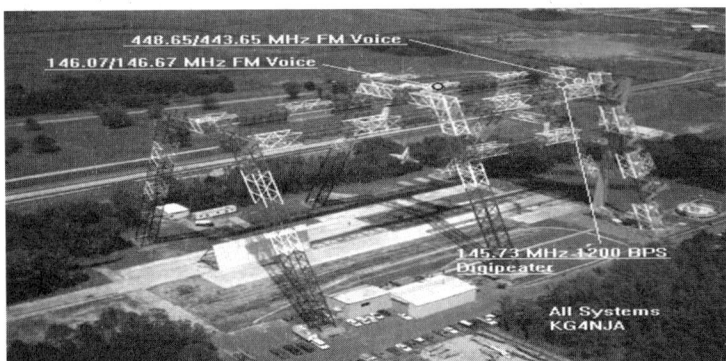

Once NASA's secret location was uncovered, it was not that difficult to collect enough information to verify this facility was indeed being used to fake the American Moon Landings. It was then just a matter of time before determining how it was being used and solving the rest of the Moon landing photography trickery.

These next photos were taken at the time the American Government claimed to have sent the astronauts to the Moon. This was back in the 1960's and 1970's. These images confirmed this facility was fully operational at the time and was capable of faking the Apollo Moon landings. All NASA had to do was say those magical words "Take one, action", and the astronauts appear to be on the Moon.

The Sliding Background Moon Scenery Was Produced Under the Langley Research Crane

Underneath NASA's Moon Landing Gravity Simulator Crane, was a large white A-frame support structure. It looked a lot like a drive-in movie screen. If this were a real Impact Dynamics Research Facility for testing the effects of crashing an aircraft, this wall would have provided protection against flying objects as the plane crashed into the surface.

However, when NASA used this facility to produce the American Moon Landing film footage, this white A-frame wall served a much different purpose. This white A-frame was used to support large 3 dimensional plaster paintings of different backdrop sceneries for the different Apollo Moon landing sites.

NASA's photographers lined up four of these walls in a row. This allowed them to place Moon mountain scenery paintings to create the illusion of a larger Moon landing site.

This movie stage platform design enabled the stage crew to change the platform for different scenes within hours. This is similar to a stage crew working for a stadium, converting a football field to an ice hockey rink overnight. This explained why the same background was seen at several different Moon landing sites. The stage crew mixed up the ground scenery over night, and since it was a live broadcast for the next day, they had no other choice, but to use the scenery even though the background was still mixed up.

This also explained, why during the Apollo Moon landing mission years, NASA hired so many photographers and artists; they certainly did not take them to the Moon to work on the pictures. If the Moon landing photography were real, they would not have needed all the artists and photographers.

Below is an illustration of how the movie stage was designed to function. Behind the artificial Moon surface, was the special A-frame wall, which the background scenery paintings were placed on. The picture on the right shows has how the movie studio could have looked with the artificial Moon surface and backdrop Moon surface in place.

Sliding Scenery

Support Wall for scenery paintings

There you have it. Within minutes, the astronauts could be seen conducting live, many different activities on the Moon at any given place or time. While everyone assumed the astronauts were sleeping on the ship, during the night stage crew workers was changing the scenery. Then, bright and early the next morning, the astronauts appeared as if they had traveled to a totally different Moon landing site.

Moon Landing Movie Stage Is Complete

In this next visual rendering, taken at the time of the Apollo Moon Landings, we see this elaborate movie studio is fully operational. Examined under a microscope, it's evident that the large artificial Moon background paintings were mounted upright onto the A-Frame support structure. Also, the artificial Moon surface platform was set up. Hence, during the alleged Apollo Moon landings, this proves that the platform was an identical match to the studio. Additionally, it was completely operational.

This image proves that NASA constructed an identical matching Moon landing site for the Apollo Moon Landing Missions and was using it at the time. Is that exciting or what? Visual confirmation on a theory! WOW! This confirmed the Moon landing action film, with the same background scenery, could have easily been filmed on Earth.

NASA's Elaborate Moon Landing

In the above picture, we see one of the landscapes for the Apollo Moon landing site was propped up behind the artificial stage platform. The background paintings appear to match the Rio Grande, New Mexico scenery found here on Earth. This is where the Apollo 17 Training Missions took place. Interesting, why would this artificial Moon scenery, under the Langley Crane, match the Apollo 16 Moon Landing Site? Is this discovery of an identical Moon scenery one of the most obvious clues to the fake Moon landings?

Moon Scenery Painting Discovered In A Warehouse:

As described in the discovery of the actual large Moon mounting background painting, NASA used this facility to store this work of art in the hanger at the Langley Research Center. This visual rendering was taken during the alleged Apollo Moon Landing Missions, which were supposed to have taken place. This can be verified by the NASA photograph below of the inside of one of the warehouses NASA claimed was taken during the Apollo mission era. The testing Apollo capsule, seen in the inflated helium parachute hovering above it, can also be used to determine the dates.

NASA ID#

Superimposing the Images Provided Additional Evidence To NASA's fake Moon Landings

Like the Apollo 17 Mission images, superimposed earlier in this chapter, these Apollo 15 Moon Landing images also show the astronauts' ship at two different locations.

Nevertheless, the Moon landing photo on the right, also provided another major clue that confirmed this footage was faked using the described sliding scenery method. Moreover, only the suspicious trained eye is likely to ever recognize mistakes NASA's artists made in this Moon Landing scenery.

NASA must have assumed no one on the planet would ever figure out exactly how they managed to fake the Moon Landing film footage. Therefore, they were not overly concerned about releasing these doctored-up photos. Toward the end of the Apollo Moon Landing Project, when the public began to lose interest, NASA's artists became lazy and neglected to do a thorough job in touching up the photos. In hopes of the landings being uncovered, there was also the possibility that NASA purposely started about leave clues about the phony Moon landings.

Anyone of those scenarios would explain why the Moon landing picture, on the right, showed the levitation cables that lowered the LM Spacecraft onto the Moon's surface platform. Furthermore, the cables were still attached to the ship's corner post. Below is an illustration of where the levitation cables were attached, and how they would have been used.

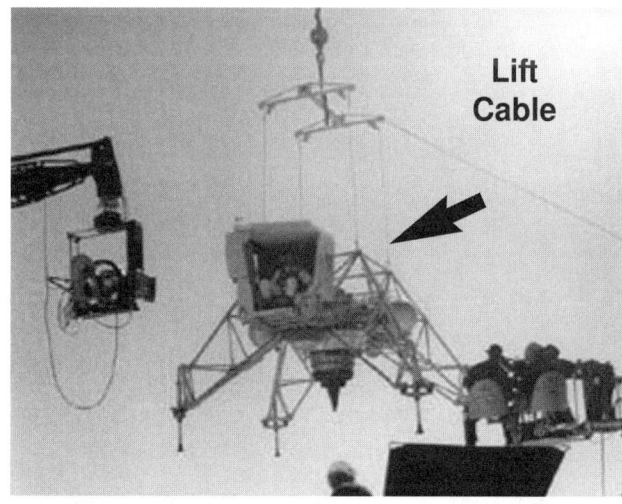

NASA ID# a17.1410123_dmh.

As you can see, there is plenty of evidence indicating NASA's Apollo Moon Landings were phony. If they were not, then why doesn't NASA just explain to the American people openly what went on here? After all, they have paid for the mission. Many people would be satisfied if just one piece of NASA's evidence of proof showed that the astronauts went to the Moon. If NASA is telling the truth, then why can't they just bring out a team of their best scientists and engineers and refute the allegation described in this book? The reason they refuse to respond is because the Moon landings were bogus. Regardless of how big NASA's budget is, they can't change the laws of physics.

Below are a few of NASA's Apollo 15 panorama shots. The landing site changed it's appearance so many times that it was unimaginable how anyone could consider them to be authentic. In determining what the Apollo 15 Landing sites were going to look like, NASA must have released the images that the artists were originally working on.

The Creation Of the Apollo 15 Landing Site Code Named "Project Genesis".

a15pan1473840

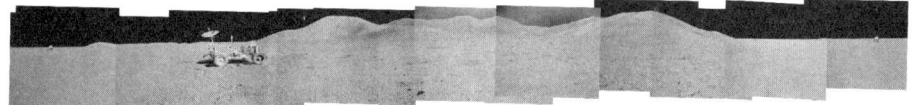

a15pan175859

a15pan1652114

a15pan161856

a15pan1441417

a15pan1650509

Apollo 15 "Project Genesis" Continued

If the panorama visual images were taken by the cameras on the Moon Rover, and there was only one rover, then these pictures could never have been taken on the Moon since the Moon Rover Vehicle, and the attached cameras were also in the illustrations?

a15pan1231715

a15pan1452812

Confirmation the Apollo 15 LM Spacecraft disappeared.

The next set of wide angle view images of the Apollo 15 landing site once again show the ship was in the pictures and then disappeared.

a15pan14480225

a15pan1650509

CHAPTER 13

Astronauts Training Locations on Earth Match Landing Sites on the Moon

Chapter 13
The Earth Locations That Match Landing Sites on the Moon

Finding the locations on earth where the Apollo moon landings took place, at first, seemed like an impossible task. There are thousands of NASA moon landing photos, and to match just one of them with an Earth location, would require a search of potentially billions of different places.

White Sands Test Facility, New Mexico

To make matters worse, evidence suggested NASA had been scattering around the landing sites here on Earth, making it even more difficult to find where the moon landings took place. NASA was unquestionably attempting to make it impossible for earthlings to ever find all the locations. They clearly didn't want people to make the connection between the earth and moon scenery. NASA's idea of scattering around the landing sites was ingenious. It was like trying to find one particular piece of sand at the bottom of the ocean. Just as we were about to give up and stop trying to find the earth locations, we had a miracle break through.

Unfortunately for NASA, when we presented the challenge to the WhizKid Mitchell (age 14), he found the challenge to be very easy. It was the most remarkable thing I have ever seen a human do.

Mitchell suggested there was a simple answer to why no one could find the matching locations of the landing sites here on earth. Mitchell said everyone was analyzing the evidence using typical 3-dimensional thinking patterns. Mitchell believed if NASA's evidence was analyzed differently, the earth locations could easily be found. Unfortunately, no one had any idea what he was talking about.

Next, Mitchell asked if he could be left alone to look over all the evidence, and he would have the answer by the next morning. Naturally, anyone hearing such a notion from a 14-year old would be skeptical, but this WhizKid Mitchell has much analytical experience already. Mitchell was recorded on a home video reciting the ABC at 9 months old and could name and recognize all the basic color groups. He has been fixing computers since he was 7 years old, and was hired by a Dot COM company as a part-time network administrator at the age of 11. Mitchell has been an avid day trader since he was 13. In addition he is developing his own Stock Trading System, which he claims, when perfected, he will make Warn Buffet look like an amateur.

That next morning Mitchell showed up yelling, "I got it! I got it!", "I found all the places NASA filmed the moon landings here on earth. Mitchell says, "It's all so simple." Poor Mitchell looked like he had not gotten any sleep that night. Mitchell then said, "NASA left the biggest clue and it reveals everything." Mitchell's theory suggested instead of looking at all the photos and trying to match them up, he developed a numerical program that could decipher the numbering sequence of NASA's entire photo cataloging system.

Now that's a different way of thinking at its best. That boy has to be one of the smartest kids on the planet because he came up with this discovery.

With a few improvements, not only did Mitchell's program decipher the exact matching locations of the moon landing sites here on earth, it uncovered so much more. You will see more in the chapters to come.

WhizKid Mitchell's program matched the scenery the Apollo astronauts used as moon training sites here on earth, to the scenery they supposedly discovered later on the moon. Again, the scenery where the American Astronauts had done their mission training on earth matched the scenery found on the moon.

Mitchell's program shows NASA's photographers simply took the background scenery from these mission-training locations. After that they combined them with the simulation pictures taken at KSC and LRC. Last but not least, NASA published them as genuine photos taken on the Moon. Now, if you find that hard to believe, just wait until you see the proof.

There is an abundance of evidence revealing that between 1967 and 1971, the Apollo astronauts traveled with Mr. Von Braun and a team of photographers and film producers to desert and volcano locations around the United States.

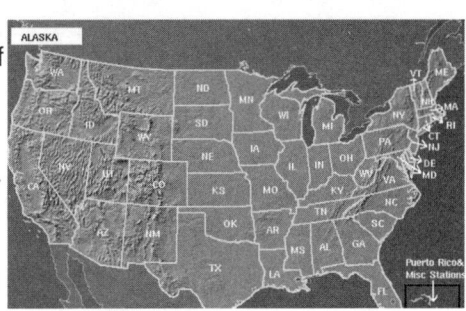

These journeys were disguised as surface terrain training missions for the astronaut's upcoming trips to the moon. However, the primary purpose of these Apollo training missions was to obtain photos of the astronauts around surface terrain that could be used as moon surface scenery.

Each Apollo astronaut training location was chosen for it special features like craters with topography scenery that was expected to have been seen on the moon. There are many clues that NASA filmed the moon landing on earth. For instance in the Astronaut's moon landing mission transcripts the astronauts make reference to being at these training location rather than on the moon. For example, Astronaut H.H. Schmitt accidentally called one of the craters on the moon in the Haemus Mountains west of Sulpicius Gallus. "Little Dan". That just happens to be the name of a 600-meter crater in the Nevada Desert here on earth.

The pictures below illustrate how easy it was for the astronauts to have mistaken the name of the earth locations to the matching moon locations. It would have required a great deal of concentration on the astronauts part to correctly apply a Moon landing site description to the earth location where they were doing the actual filming.

In these NASA pictures below the Apollo Astronaut can be seen comparing the Earth locations that would later be used for their upcoming Moon landing sites.

Astronauts studying Earth training location to be used for their upcoming Moon Landing hoax sites

NASA ID# 10075922

The Earth Locations Where the Moon Landings Were Filmed

Originally, everyone was concerned that if we provided everyone with specific directions to the exact locations of the fake Apollo moon landings. People would lose interest. The mystery would be solved and the excitement would be lost because there would be no one left on the planet to debate if the moon landings were real or not. Everyone would know the answer.

Apollo Astronauts NASA ID# ap15-s71-23773
Rio Grande Gorge, New Mexico - US

Close up of
Background Mountains

Oh well, what the heck, this American Moon Landing hoax has gone on long enough. Lets put an end to the moon landing deception before the poor American taxpayers go bankrupt. While their government tries to continue making up phony evidence to support their ridiculous claims. Here are most of the Apollo moon-landing sites hear on earth. The Grand Canyon, the Big Bend Area of West Texas, the volcano fields near Flagstaff, Arizona, Cimarron, New Mexico, The White Sand Training, and Area 19 "Shundahai" Nevada Test Site to name a few.

Some of the locations will not be revealed in this book, to still leave some mystery behind the Moon landing hoax.

If you want to take a trip to the moon, and see it just as the Apollo astronauts did back in the 1960's and 1970's, try visiting anyone of these location NASA used for most of the moon background mountain scenery. Here, you'll can see most of moon mountain ridge found in the pictures NASA claimed to have taken on the moon.

Along the Rio Grande Gorge, where NASA staged a lot of the phony landing scenery, is found on the broad valley between the Sangre de Cristo and San Juan ranges of New Mexico. From there, you can recognize and identify the matching features of the Apollo 15, 16, and 17 moon-landing scenery all around you. The Rio Grande Mountains, twenty miles to the north, are an identical match in distance, shape, and height to the Hadley Delta Mountains on the moon. Now, look to the West, and you'll see the muddy Rio Chama which was used as the Hadley Hille Gorge, alleged to be on the moon.

Wait a minute, just because the scenery at "Rio Grande Gorge" matches, almost identically to the background scenery in the Apollo moon pictures; how can we be positive this site has any connection to the American Moon landings? Could it just be a coincidence that NASA's Astronaut training sites on Earth just happen to resemble most of the landing sites found in different places on the moon?

a15dmh1231715

Wild River Rec Area
Rio Gransa Gorge. NM

Apollo Astronauts NASA ID# ap15-s71-23773
Rio Grande Gorge, NM

For those of you that are still not one hundred percent convinced the NASA moon landing scenery is fake, check out this next set of pictures. Here, we see the Apollo 15 Astronauts Scott and Irwin, as they overlook the suspected moon-landing site here on earth just before they supposedly make that dangerous mission to the moon.

These pictures provide a close up of the scenery that matches the Apollo 15 moon landing sites. They are complete with the Apennine Front at Hadley Delta, and the Hadley Rille Gorge.

Common sense would suggest it would be impossible for the Apollo 15 Astronauts to be standing at a location here on earth that matches the same place they are soon going to visit on the moon. If you're still not convinced, these matching locations are for real. If you think the photography is not true, then take a vacation someday to this place called "Rio Grande Gorge". As previously stated it's located in New Mexico. The view is spectacular.

By the time you've received this book, most likely already half the concession stand businesses from Roswell, New Mexico will have moved to this location. By now, there is probably a new "Moon Mountain Ridge" theme park complete with food stands and rides for the kids.

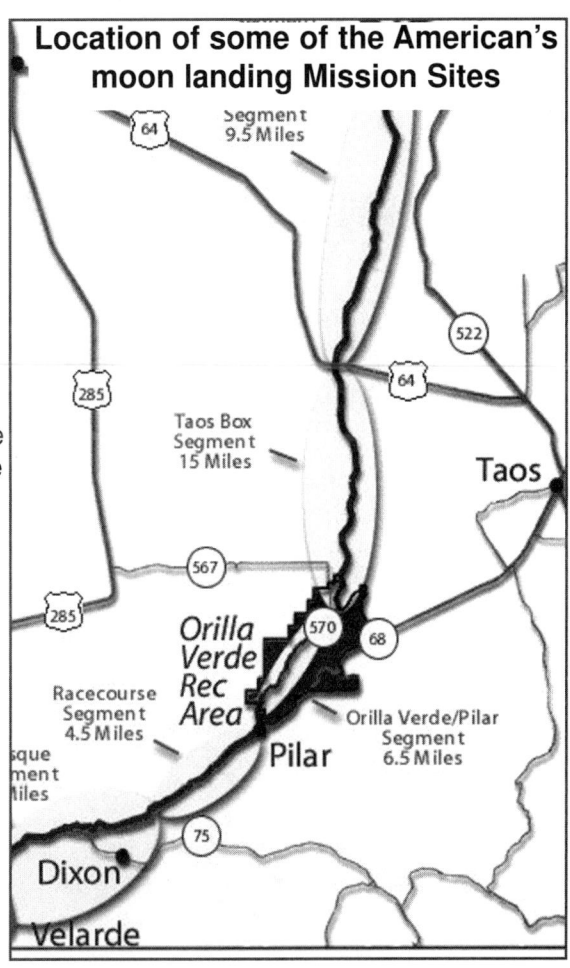

The Pancake Range area of south-central Nevada was also used as a location for the Apollo 15 moon landing missions.

The top photo on the left is the matching background scenery with Apollo Astronauts Eugene A. Cernan and Harrison H. Schmitt is the foreground. According to Mitchell's numbering formula, the hill in this picture on the right is the same hill used in the Apollo 15 moon landing scenes. As you can see, the Earth landscaping is an identical match to the alleged moon landing scenery.

Apollo 15 Landing Site

NASA ID# 10075923

EARTH

NASA ID# 10075922

ap15.1473840

MOON

The purpose of these NASA training missions was to size up the astronauts with various realistic looking background scenery for NASA's moon hoax. Then, the NASA photographers would insert the selected background scenery into the mission photos filmed at the Kennedy and Langley Space Centers. It was that simple! This evidence will be shown in complete detail in the following chapters.

NASA ID# ap15-71-H-646HR

NASA ID# as11-40-5954

The Rio Grande Gorge between Pilar and Taos, New Mexico - Apollo 15 Mission

Here, we see the NASA photographers sizing up the astronauts with a large rock for a later moon-landing scene. Once again, we have another match according to NASA's photo numbering system. It's obvious this is the same location and rock formation, and the hill is sloped with the same pitch. The rock is about the same size and also has a flat top, very strange indeed. What's even more suspicious about these two pictures is; the rocks near the base of the large moon rock are almost an identical match the ones beside the rock on earth. NASA's artists go to work filling in the artificial moon surface, made up of sand and Volcano ash, and there you have it.

NASA ID# as15-s71-39725

NASA ID# a17slop

Now, we can try to believe that out of six billion people living on planet earth at the time, these two guys stubble upon this very unusual rock formation. A few months later, these same two guys, travel to the moon and find the same type of unusual rock formation.

Let me tell you, I have a pretty good understanding of numbers. But I could not even begin to figure out the odds of this scenario taking place. Without doubt, the odds of these astronauts finding just one similar match site on the moon would be such crazy number like "one out of a Gazillion". That's not even considering these astronauts were filmed in at least two other locations on earth that are almost an identical match to the scenery, the astronauts supposedly encountered later on the moon.

Can you imagine if the Los Vegas Casino's were giving odds on these NASA moon landing photo's being real. If you were to bet just one dollar on NASA's moon landing photos being real and NASA could some how prove they were real. Your one-dollar bet would make you an instant Gazilionair. Congratulations! With your winnings, you now make Bill Gates look like a welfare case. Unfortunately, after this book is published the Casino's are never going to take bets on NASA's moon landing photos being real. Even at a Gazillion to one. But if they were, I surely would not waste my dollar.

The point trying to be made here is; there is no real mystery as to why the Apollo Astronauts Training Scenery is almost and identical match the scenery the astronauts supposedly discovered on the moon. NASA artists simply used the astronauts training scenery, doctored it up a little bit and used it for the moon landing scenery.

When you think about it. The odds of NASA using the matching Earth scenery in the training pictures, for the moon scenery, is less then a fifty-fifty. NASA either used the matching earth sceneries to fake the moon scenery or they did not. The funny thing is even with a less then fifty-fifty chance NASA faked the moon landing photography, the casinos would never be willing to take any bets that the moon landings were real. However I have no problem seeing the Casinos taking bets Gazillion to one that NASA faked their moon-landing photo.

Pancake Range Area of South-Central Nevada. - Apollo 17 Mission.

The picture on the left is of the Apollo 17 Astronauts Eugene A. Cernan (right), and Astronaut Harrison H. Schmitt on the left. They are at the Pancake Range area of South-Central Nevada. Here supposedly they are conducting extravehicular training exercises. The two pictures on the right, NASA claims to have taken them on the moon during the Apollo 17 extravehicular mission at the Taurus-Littrow landing site (station 4). Notice the earth training location and the moon landscaping has the identical background scenery.

10075922   as17-22199-201   10075963

These matching sceneries are clearly irrefutable proof that NASA faked the moon landings. No matter how good NASA's scientists were, it would have been impossible for them to predict the lunarscape at Station 4, and then match it to a location here on earth for training purposes.

Coso Range Area, near Ridgecrest, California.- Apollo 15 - Station 2

Here is an example of another moon landing site that matches one of NASA's training mission area's here on earth. Using the photo numbering sequence formula to match this landing site, it was determined that some of the background scenery came from the Coso Range area, near Ridgecrest, California.

Below is a part of the Apollo 15 Station 2 landing site background.

After combining the matching earth location photo to the moon background, it was an identical match. It was visible from the angle taken in the NASA picture 10075798. From this low line photo shot angle, the hill above the man on the far right, in picture 10075799 would have disappeared making an exact match to the moon landing site. The matching photos NASA claims were taken on the moon during the Apollo 15 mission, that ridiculous.

Area 19 - Nevada Desert Scenery and Craters

When considering the fact that all of the Apollo Astronauts training locations resemble the alleged astronauts landing site on the moon in one way or another, it becomes impossible to calculate the odds of this being a coincidence. NASA had to have used these training sites to stage their moon landing hoax. For example take the Area 19 location located in the Nevada Desert, it was scenery was ideal for staging the moon landing hoax. This location at the time was off limits to the public and the huge bomb craters looked like they would have been found on the moon.

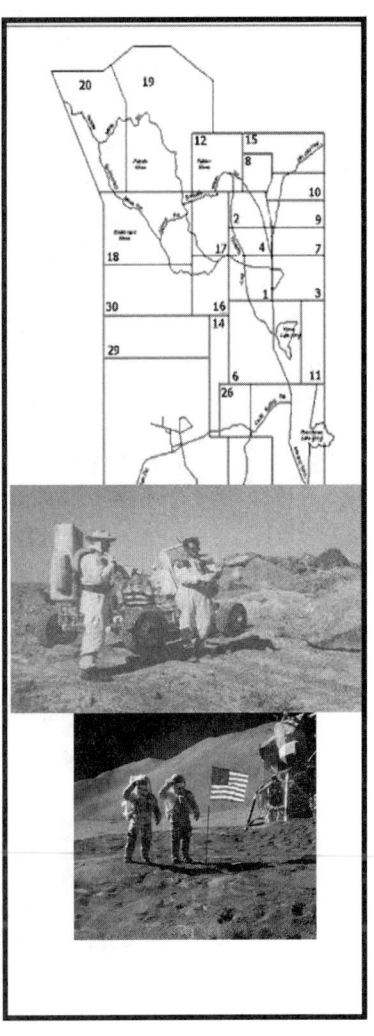

Area 19 "Little Danny" Crater matches Haemus crater on Moon

Below are the two almost identical craters. The crater on the left was supposedly on the moon and the one on the right is a matching crater on Earth. It's easy to understand why Astronaut H.H. Schmitt accidentally called one of the craters on the moon the "Little Danny". This is the name of a 600-meter crater found at one of the locations on Earth in the Nevada desert believed to have been used by NASA for filming the Apollo Moon landings. These two craters help illustrate how little NASA artists change the earth crater's appearance for their moon hoax.

Astronauts Moon Landing Site

Astronauts Training Site on Earth.

The Kilauea, Hawaii Site-Apollo Mission and Moon Surface Soil

The photo on the left with NASA ID #10075463 was entered into the photo filing formula, and the rest of the rock pictures of the Apollo missions came up. If you look from the right of the picture of the astronaut posing for a moon shot, the moon rock matches the rock the astronauts are taking a photo in front of.

Apollo Astronauts Lovell and Haise during training at Hawaii Description

Apollo 16 - 10075844
Moon near South Ray crater

NASA ID# 10075954

as16-108-17701

10075628

10075626

10075843

The remaining rock photos also came up as matching this earth location Kilauea, Hawaii. If you look closely at the NASA picture #10075463, you will notice that this Kilauea, Hawaii, surface composition matches the soil used for the moon surface films and photos. That is because volcanic ash is what NASA often used as the top layer on their movie stage platform. That's right, NASA used the volcano ash from Kilauea, Hawaii mixed with sand as the moon surface. As more and more matching locations were found, it was obvious NASA's filmmakers were intentionally scattering the filming locations around, so no one would ever be able to pinpoint the exact locations of all the sites here on earth.

The Rio Grande Gorge between Pilar and Taos, New Mexico is one of the Apollo 15 Moon Landing sites. Refereed to as "Moon Mountain Ridge"

Here, again, the Apollo astronauts are at another location taking pictures in front of the background scenery that is going to be used for the Apollo moon-landing hoax. As usual, NASA claims this was only geological excavation training for the Apollo astronauts. Yet again, just like the other training missions, the background scenery seen behind the astronauts matches the scenery on the moon. Below are a few samples of the moon pictures that came up as matches to the Taos, New Mexico area using WhizKid Mitchell's formula for the NASA pictures.

One of Apollo 15 Landing Sites

NASA ID# 10075983

Pinacate Volcanic Area of Northwestern Sonora, Mexico. -Apollo 14 Mission

When Mitchell's earth and moon location matching formula linked these next two pictures, we thought his program had a small glitch in it. The background scenery in these two pictures are completely different, yet according to Mitchell's system, the two pictures are from the same place. Visibly, the Apollo Lunar Surface Experiments Package (ALSEP) in the earth picture is on the edge of the Pinacate volcano, and in the Apollo 14 moon picture on the right. It has a different flat surface behind it. How could this site be considered a match?

10075577

10075627

Wait a minute! Look how the rocks at the base of the ALSEP Tool Cart match in both pictures. The rocks are the same size, shape, and arrangement in both pictures. Noticeably, it would have been impossible for NASA to predict the size and exact formation of the rocks the astronauts would later find on the moon without ever being there. How could NASA expect anyone to believe the ALSEP Tool Cart was brought to the moon and placed in an identical rock formation? These photos are undoubtedly both from the Pinacate Volcano area here on earth.

Immediately after making the rock connection, it was obvious the moon picture was another composite photo. Once again, NASA's photographers simply cut and pasted pieces from other pictures around the ALSEP photo taken here on earth. They did this to create an entirely new background, so one would make the connection. This is a very common photography trick and was used heavily by NASA. Below are more images with matching number sequences. These portraits confirm that they were used to compose the moon pictures.

NASA ID# a15pan1223653gmh

as17-68-9486

NASA ID# 10075637

10075584

he desert area at NASA's White Sands Test Facility, New Mexico.

The White Sands Testing Facility was also used for segments of the moon landing film footage. And was heavily used for the Apollo 17 moon landing area scenery.

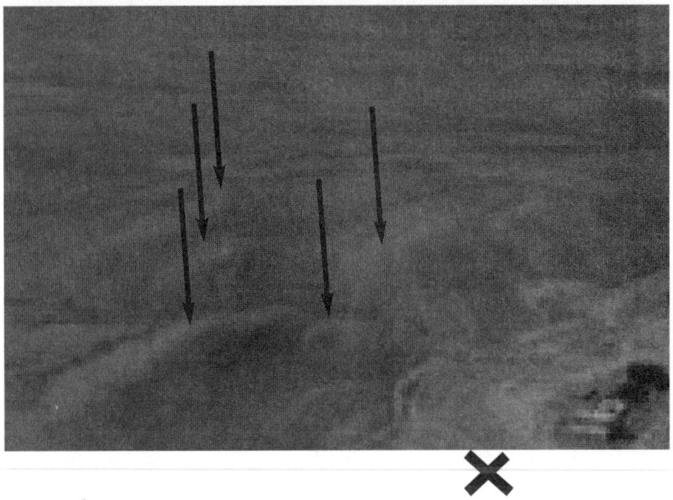

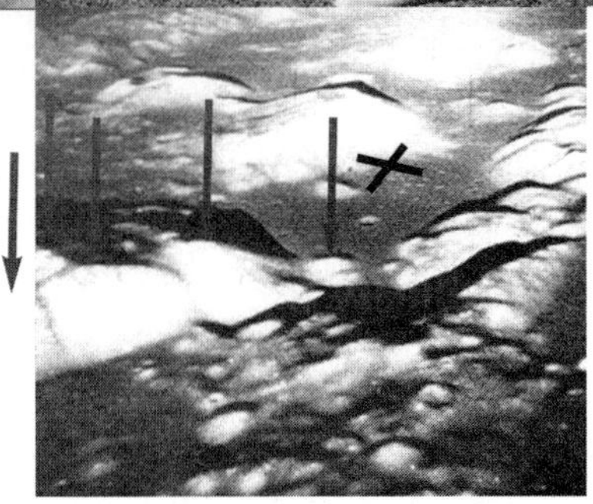

The Rover Vehicle Testing Site Matches Moon Location

Another earth scenery that matches the scenery found on the moon. It's funny how the hill in the background of this rover testing facility, matches the hill on the moon where the astronauts are also testing the rover vehicle on earth, in his space suit. Is NASA leaving these clues all over the place hoping that someday people of the future will figure out the Moon landings were a hoax? Most likely soon after this book is released that very question will be answered.

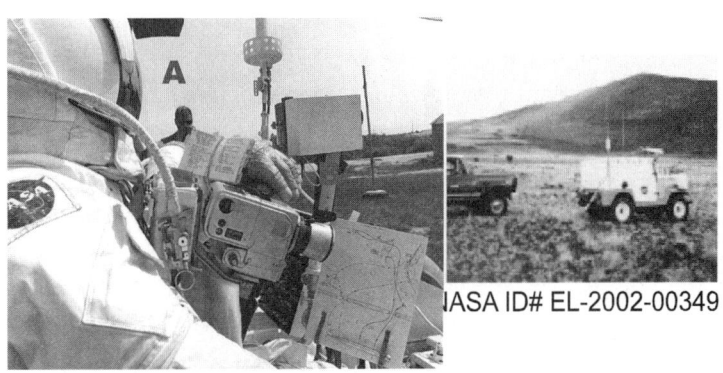

NASA ID# EL-2002-00349

The WhizKids Experiment #7 "

The WhizKids suggested organizing a massive ecological dig at various Earth locations, suspected of being used as Apollo moon landing sites.

If a large number of people immediately began conducting a search on these areas for clues left behind by NASA's filmmakers and the Apollo Astronauts, they would definitely find clues that NASA was there quickly.

The WhizKid's are right; a quick massive search would be helpful before NASA starts to seal off these earth locations in an attempt to conceal their moon-landing hoax. Although, quarantine these areas at this point will only spark a public outcry. The American Government has already been exposed and NASA is better off to take their chances and hope nobody finds anything.

The WhizKid's Field Trip Fund-raiser:

The Internet WhizKids setup a fund-raiser to pay for field trips to the Apollo moon-landing sites for financially disadvantaged students.

Their goal is to eventually insure each region on the planet has at least one eyewitness who can relay their experience and provide a confirmation of the moon-landing hoax to their schoolmates and teachers.

The WhizKids found it sad to see so many children on Planet Earth are still held back from their true potential. They are forced to make decisions based on lies and deceptions, which are forced onto them by the planet's super powers. Whether it's the moon-landing hoax or the truth about Earth Ozone depletion being caused by nuclear weapons testing in outer space, the lying needs to stop. Evidently, there is already plenty of evidence proving the nation's leaders have a total disregard for the planet's environment. This lack of concern is already creating global devastation. It is not only affecting human beings existence, but also all earth's life-forms.

There is without a doubt enough evidence contained in this book that anyone can now confirm scientifically that NASA's claims of having technology capable of human travel to the moon and beyond is not true. It's not uncommon to see a large section of earth's population put their faith in one super power. The entire future of the planet is being altered by this power.

Unfortunately, the American Moon landing hoax was different. The deliberate falsification of scientific evidence is a serious matter. The planet's direction of technological advancement has been severely changed at the most dangerous time of the planets history. Presently, we are experiencing the beginning stages of industrialization. The moon-landing hoax has caused a destructive shift in the natural course of technological advancement. Because of the misinformed belief that deep space travel is just a few years away, many humans believe that global devastation is of little concern. Further more many people believe they will soon be able to occupy other planets. Unfortunately, because of earths radiation belts, no one will be going anywhere soon enough. Currently, this is the only planet you're getting, so if you don't take care of it you will perish.

Perhaps the people of the future will look back at this Moon Landing Hoax as a critical turning point. In the future, might people look back in history and realize the hoax changed the natural time continuum for earth's technological advancements and put humans at risk of extinction. Think about it; thanks to NASA and the American government, the planets primary focus is on primitive nuclear power, which is very dangerous. Would humans, if not miss-informed by the American Government about space travel capabilities, begun to focus on safer and more powerful alternative energy sources, such as particle physics and harnessing magnetic flux line power? Would humans have eliminated the use of primitive nuclear power before destroying their environment? If the scientific community were presented the true nature of NASA's scientific findings much sooner, would that have stopped the destructive testing of nuclear bombs?

If you are, or if you know of a disadvantaged student that would like to sign up for a chance to get an all expenses paid trip to where the Apollo moon landings were filmed, please register at the WhizKids web site at www.whizkids.tv. If you would like to make a donation and help send an economically disadvantaged student to "Moon Mountain Ridge" to help ensure the truth is told. You can also make a donation by going to the WhizKids Project Web Page at www.whizkids.tv or www.moonbloopers.com. You'll help ensure that this type scientific deception does not spread to new super powers of the future.

The Earth Locations That Match Landing Sites on the Moon

Example of photographers having difficulty choosing how they wanted certain scenes to look!

This Apollo 17 moon-landing scenery has been a little troubling for the NASA site designers. It seems as if they were not sure how they wanted it to look and tried several different styles.

a17pan1464906

a17.1653338

a17pan1645954

a17.1662658

a17pan1445305

CHAPTER 14

How NASA Created the Moon Craters

Chapter 14
How NASA Created the Moon Craters

What about all those craters the astronauts were seen working around on the moon surface? Could NASA fake something as big as moon craters?

Here again, using WhizKid Mitchell's program, it was possible to match the moon craters to specific locations here on earth. Subsequently, after matching the Earth crater photos to the moon craters, it was possible to determine the various methods NASA's photographers used to create the realistic moon crater scenes. Starting with the artificial craters, this chapter will provide an explanation for each method believed to have been used to create the moon craters.

NASA Used Man-Made Moon Craters

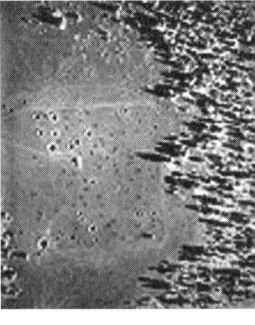

One location, where NASA used artificial moon craters, was at the Cinder Lake Crater Simulator in Arizona. This site was secretly named "Grover". Like many other apollo astronaut training simulators, the scenery appeared to match the scenery later found on the moon. Is that even possible? The astronauts training locations somehow mysteriously matched the moon surface.

The next representation on the left shows NASA employees building man made craters at the Cinder Lake crater simulator. Here, we see firsthand NASA employees building ARTIFICIAL craters. Notice how the man made crater is an identical match to the crater supposedly filmed on the moon by the astronauts.

NASA photographers did not even alter the appearance of the crater, or hill, in the background. Under a microscope, the piles of brown soil, dug up at the last minute, were left in their place. The same dirt pile was visible in the matching moon picture to the right. NASA either made a mistake, or was leaving everyone clues in hopes that one day they would be found.

LEFT Crater

RIGHT Crater

Cinder Lake crater simulator in Arizona.

Next, is another perfect example of how craters at Cinder Lake matched those seen on the moon. The visual image on the left shows two Apollo 15 crewmembers inspecting one of the craters right before their alleged trip to the moon. Now look at the picture the astronauts claimed to have taken on the moon surface. When you take away the rocks placed next to the moon crater, the holes look identical. Could this simply be another one of NASA's so called unexplainable moon anomalies? I don't think so. The training simulator crater; was taken after NASA's stage crew had already started adding the artificial moon dust.

LEFT Crater　　NASA ID# 20117349
RIGHT Crater

14-2

Matching Man-Made Craters Found In Texas

To avoid detection in building the artificial Moon surface, NASA used several locations to film here on Earth. In the next set of NASA photos, the top picture with the astronaut standing next to the crater was taken at the Johnson Space Center in Houston, Texas.

Under close examination, this man-made crater is almost identical to a Moon crater below. NASA is correct in some cases, the craters are not identical matches, however that is only because the artificial Moon soil and artist touch-up has not been added yet. Even if these earth and moon craters in some instances are not a match in every little detail, the odds of man going to the Moon and finding that almost every crater is almost identical in size and shape to the craters used for training here on Earth, would be something like 1 in a 1,000,000,000,000.

More Man Made Craters found on the Moon:

Below is another example of a man-made crater being used as Moon scenery during a different Apollo Moon landing mission. What was NASA thinking?

Combined Uses of the Man Made Craters

As seen before, to distort distance, NASA photographers cut and pasted pictures, and the moon craters are the same type of deception. During the Apollo 15 Moon landing mission, the moon crater on the right was alleged to have been discovered, however it is nothing more than a composite picture of the Houston, Texas crater on the left. Here is how they did it. First, the crew filled in the moon dust surface. After that, NASA photographers took many pictures, at different camera angles, and pasted them together.

NASA Photo # a16pol1665800-2
Man-made crater

Composite Craters:

These next two images show how NASA took the cutting and pasting of craters one step further. While creating the Apollo 15 moon landing scenes, the top picture, was supposedly Station 4. Notice how the crater is halfway filled in. The lower picture, is the station 6 landing site. The site was supposedly over 50 kilometers to the left of the mountains in the background. Besides the fact that these two pictures are almost identical scenes, notice how the crater has been completely filled in on the picture to the right. Under close examination, it is very clear that these craters had been filled in because there are hundreds of little camera crosshairs visible in the crater area. That's why, this proves pieces from other photos were used to fill the crater.

NASA doctored these next two photos by filling in the crater:

There are many more examples of NASA photographers altering the shape of their man made craters with pieces from other photos. For a second time, they did this to create a totally different moon scenery. This is nothing new. This type of fraud has been around since the first films were made, and is referred to as composite photography.

NASA gets a bang for their buck, when it comes to the Houston Crater

The large simulator crater in Houston, Texas was used in the Apollo 14, 15, 16, and 17 Moon landing scenery. One would assume the Houston location was a very popular place for filming the moon landing hoax, since many of the Apollo astronauts live in the area. Certainly commuting to the other Moon landing sites here on Earth would have been annoying for the astronauts.

ID# 04191bbo

NASA PHOTO ID#:a16.plum_pan

One startling discovery was made about the NASA Apollo 16 mission. In this operation, American astronauts worked around a crater that was spread out over many kilometers. After cross referencing the different site photos, evidence suggested that the sites were filmed in front of one small crater, which was roughly 70 feet wide. The Houston, Texas Simulator Crater is shown above. The rough edges, gaps, and pockets between boulders were simply smoothed out with a mixture of sand and charcoal ash. To illustrate this, look at the alleged moon crater photos below.

As seen in the NASA pictures below, this same crater was used in several scenes.

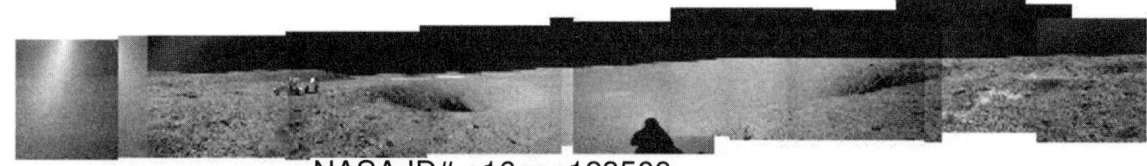

NASA ID# a16pan123508

NASA ID# a16.plum_pan

NASA ID# a16.flag_pan

Actual Earth Craters Were Used To Fake Moon Craters:

It appears not only did NASA photographers use actual earth mountain sceneries for the moon landing pictures, but they also used several craters found here on earth.

For example, during the Apollo 15 manned mission, the astronauts supposedly discovered the Hadley Rille Gorge. However, this gorge matches the Rio Grande Gorge, in New Mexico with minor artist alterations made to the gorge appearance.

Apollo 15 Mission NASA ID# a15dmh11284-8

Unmistakably, the Hadley Rille Gorge that the Apollo 15 astronauts claimed to have discovered on the moon was an identical match to the Rio Grande Gorge on earth. Likewise, the astronauts used the Rio Grande Gorge as a training location. The shadows on the right side of the earth and moon craters even match. The earth and moon canyons have the same depth, hill pitch, and curves. In addition, they have similar shadow casts. Indeed, this could not be a coincidence. They must be the same location.

These similarities confirmed the moon landing locations discovered by the WhizKids, and prompted a much broader investigation into NASA's evidence related to the Hadley Rille Gorge on the moon. Could the real purpose of NASA's outdoor Apollo training missions be to film the astronauts in realistic looking moon scenery? This would explain why a majority of the Apollo Astronaut's training photos focused on earth scenery that matched the landscape later discovered by the astronauts on the moon.

More And More Pictures Confirm the Theory Of Hadley Rille Gorge On the Moon Was Indeed the Rio Grande Gorge In New Mexico

These next figures prove that the moon landings were a complete hoax. In addition, they confirm that the landings were staged on planet earth. The wide angle view of the moon's Hadley Rille Gorge matched the view of the Rio Grande Gorge on planet earth.

For those of you that may still not be one hundred percent convinced this Rio Grande Gorge was used to fake the Hadley Rille Gorge on the moon, try to explain why the Apollo 15 Astronauts, Scott and Irwin, just happened to be standing in front of two identical looking craters. One of the canyons was on earth, and the other was on the moon.

Apollo 15 Astronauts at **Hadley Rille Gorge** on the Moon

Apollo 15 Astronauts at **Rio Grande Gorge** In New Mexico

Besides, how could the mountain scenery, on the moon, behind Hadley Rille Gorge, match the surroundings behind the Rio Grande Gorge on planet earth? Common sense would suggest that it would be impossible for the Apollo 15 Astronauts to be standing at a location here on earth that they are soon going to visit on the moon.

Remember, when these photos were obtained from NASA's archives, they were in there original form. If you're still not convinced that these matching locations are for real, or you think the photography is not authentic, I encourage you to plan a vacation someday to this place the Whizkids now call "Moon Mountains Ridge" in New Mexico, and see it for yourself. The view is spectacular.

Close Ups Of NASA's Moon Landing Photos Also Reveal Fraud

Finally, lets go back to the original close up picture provided by NASA. Close up, with a magnifying glass, it revealed that this scene was totally fabricated. To alter the earth's pictures appearance, it's clear the photographers simply used many pieces from other photos.

While examining the photos under a magnifying glass, there are many clues that NASA fabricated this moon landing scenery. It's that bad!

Notice how NASA's Hadley Rille Gorge photo on the right portrays half of the antenna on the Rover Vehicle. Did the other half grow legs? Where is it? To add to this, there are square patches with different shades. The biggest clue of all is, in this picture, there are hundreds of camera crosshairs visible on the surface.

The Whizkid's computer system, named "Amitch", calculated this scene contains 207 crosshairs from 19 different photos. They were collected from 8 other NASA moon landing sites that were spread out over a distance of 1,149 miles (1,1850 kilometers).

Using the Rio Grande Gorge, NASA photographers caused these out of place crosshairs. The artists kept a picture base, and then pasted images of artificial moon surface over that. In the world of photography, this is nothing special.

NASA ID# a15dmh11284-8

More proof composite photos were used to make moon landing scenery

If you look closely, in one fashion or another, you can see that all the other pictures were used as a group to make the moon Hadley Rille Gorge crater scenery. More evidence confirms that the moon landing photos are all 100% fakes.

14-8

CHAPTER 15

Deciphering NASA's Coded System For Hiding Their Incriminating Evidence

Step 1: Search All Possible Patterns

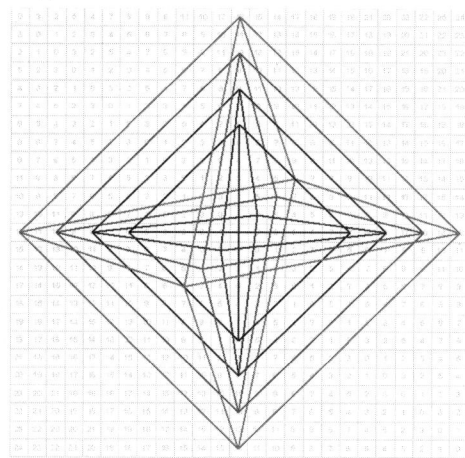

Step 2: Establish Common Link

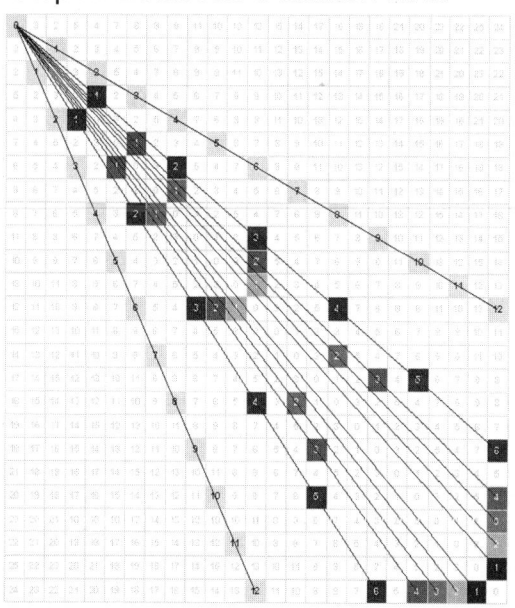

Step 3:
Decipher Messages in the Code

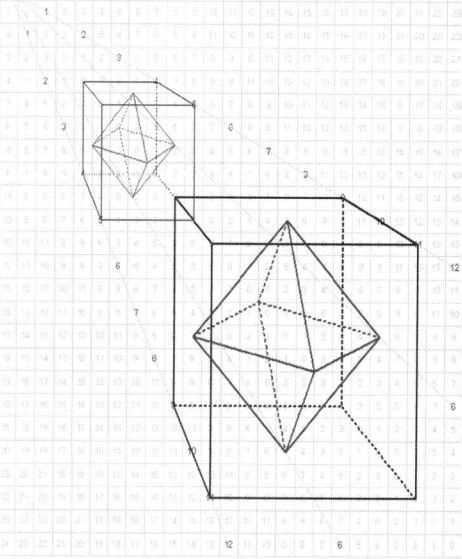

Chapter 15
Deciphering NASA's Coded System For Hiding Their Incriminating Evidence

As mentioned before, the genius Whizkid Mitchell developed a special numbering system that was able to find the locations where the Moon landing hoax was filmed on Earth. Realizing the enormous potential of Mitchell's program, I suggested he reconfigure it into a multi-developer format similar to Lionel's LINUX operating system and make it accessible via the Internet. This modification allowed an unlimited number of scientists, engineers and programmers to simultaneously work on his program and make improvements.

Modifying the Program Capabilities:

With so many brilliant minds now making improvements to the original program, it quickly caused the program to take on a totally new metamorphism. Mitchell's program eventually evolved into the "Amitch System" (Artificial Machine Intelligence Technology Computerized Hyperthinking). There were some very powerful enhancements to the "Amitch System" that enabled it to make some very startling new discoveries relating to the Moon landing hoax. Some of the discoveries are so incredible you won't believe them until you see them.

It is truly amazing these discoveries were ever made. At one time, the "Amitch System" seemed so advanced it was unlikely we could get it to work to its full potential. Amitch was going to be scrapped because every time we tried to run it the complexity of the necessary programming parameters locked up the computer. To run efficiently and decipher all of NASA's Apollo Moon landing evidence, the enhanced Whizkid's Amitch System needed the combined power of several hundred computers. Unfortunately, the decision was made that the Amitch System would have to be canceled for several years until computer technology was advanced enough to run such a sophisticated system. Finding out what went on behind closed doors at NASA during the Apollo Moon landing hoax would have to wait a view more years.

Suddenly, there was a new light at the end of the tunnel and this time it was not a train. The Internet Whiz Kids from Japan claimed they knew of a secret project their government was working on that could provide a solution to the Amitch System. Much to our surprise, the Japanese were developing a space age ultra-high-speed parallel computing super system that combines the computing power of 5,104 microprocessors. That is like combining the power of 5,104 computers working together at one time. Because of this, we are capable of running Terascale Computing up to 40 Tflops. It was so massive it had to be stored in cabinets covering the space of four tennis courts at the Earth Simulator Research and Development Center in Yokohama, Japan. The computer was secretly code named "Computenik".

Even though this was exactly what was needed to run the complex Amitch System, it was unfortunately under strict government control with 24 hour armed guards. We could not just walk up, knock on their door and ask them if we could use their secret super computer. Amazingly, the Japanese Internet Whiz Kids had a solution. They had developed a backdoor direct link via the Internet to the super computer. Previously they were using this back door computer link to play high-speed online games.

Without the Japanese Whiz kids providing the opportunity to run the Amitch System undetected 24/7 on this super computer, none of the evidence in this chapter would have been possible. For very large-scale applications development such as the Amitch system, this secret super computer was decades ahead of anything else. It included a custom language compiler that performed automatic parallelization and vectorization, such as Fortran90, C, and C++. This computer was perfect for the multi developer customization being made to the original program.

This allowed the Amitch System to perform futuristic mathematical computations. When pushing the Computenik Computer to it's limits, it would carry out over 35 trillion mathematical operations per second. The extraordinary power of the Computenik computer enables new possibilities in data analysis. This was truly a blessing in disguise since personal computers will not reach this level of computing power for another twenty plus years.

Within a very short time, Mitchell's program architecture evolved into what is now the "Amitch System". It is a totally new dimension in artificial intelligence micro processing.

This was a real breakthrough in the artificial intelligence conducting execution and selective processing detection of random data packets. With the ability easily passing 150 different command line parameters into a scanning routine with a random string of 60 characters. By detecting a binary algorithm encoding on the various data packets, it was not necessary to attempt the impossible task of establishing pre-defined linear search options. This was a key factor in deciphering the Apollo Moon Mission photo numbering system. Can you imagine what would happen if Mitchell had access to this computer to finish perfecting the Stock Trading System he is working on? WOW! The improvements made to the "Amitch System" were not to provide additional proof that the moon landings were fake because there is already insurmountable evidence supporting that theory. The goal was to search for proof that people knew about the moon hoax, and then determine what actions they took to conceal the truth. Yes, we were asking the computerized Amitch System to think like a human, and then perform advanced tasks that humans are incapable of doing themselves.

Amitch was to analyze an enormous amount of random information data packets that included non-synchronized types of data. For example, items varying from photos, videos, reports, characters, numbers, and just about anything with no predefined correlation to each other. Using an artificial computerized intelligence concept, the Amitch was to independently come up with potentially trillions of different unpredictable scenarios.

Amitch ran all the evidence through a decoding routine; however, after any connection was established, in an attempt to form a hypothesis based on the newly established connections between the random data, it would run through a series of sub routines. If Amitch had formed any type of connection, or hypothesis, they would be inserted back into random data to be searched again for a new connection. Also, for independent decoding, all the potential connections and hypotheses were sent to separate holding tanks.

To base a totally new or improved hypothesis, this search routine would run again, and try to build off of the new data looking for more in-depth connections. In an attempt to find the truth that could be presented as evidence that the moon landings were a hoax, this routine went on over and over again.

Eventually, Amitch would either disprove that theory, or come to a final conclusion to be manually cross-referenced. This feature of the Amitch System was so powerful; The Internet Whizkds used it to develop an Internet business opportunity, which allowed them the opportunity to buy all those nice things, such as new cars for their parents.

For three months, the "Amitch System" had been running with the finest computer technology imaginable and some of the brightest minds in the world. Thus, it made truly spectacular improvements. During this time, Amitch found plenty of new evidence related to the hoax, but nothing sensational. Amitch appeared to be unable to find any evidence linking NASA officials to the Moon landing hoax cover up. Unfortunately, it looked as if Amitch was unable to achieve that human like characteristic of true independent thinking and decision making based on past experiences. Then, just before this book was to be printed, we hit the mother load. Our ship had come in, and it was filled with gold. Suddenly, "Amitch System" was lighting up like a Christmas tree. It was spitting out the type of evidence we were hoping for and much more.

At first, we knew something was going on with the "Amitch System", but we had no idea what. The Japanese super computer Computenik was somehow put in a super charge mode, and the CPU speed had increased to almost 1.25 times what was perceived to be its maximum threshold.

After that, Amitch started to establish the strangest pattern in the numbering sequence of all the Apollo evidence. The super computer's CPU speed made another jump to 2.35 times the maximum threshold, and Amitch started forming bits and pieces of different theories. The CPU speed increased again to 2.70, and, all of a sudden, Amitch started pinging our office computer as if was trying to establish a link. It was looking for an I/O port to establish a link. The Amitch System was attempting to establish a link and request that we accept its transmission of data findings. This was not supposed to happen and made no sense. The Amitch protocol (procedure) was for us to retrieve all of the data stored in holding tanks. These tanks were located on the Japanese Yokoham Super Computer. Amitch was never designed to contact us, or transmit data. Amitch continued to hack into our system for several minutes while asking to complete its mission and transmit its findings. We did not respond because we thought it was Japanese engineers trying to pinpoint our location.

Then several minutes later, the Computenik processing speed increased again. Furthermore, Amitch somehow established a direct link to our computer. Amitch began configuring a new file on our desktop. At first, we assumed someone was modifying the Amitch system on the fly, but then we realized, at that speed, it could only be the Amitch System itself making these changes. Had the Amitch Program finally achieved true artificial intelligence and somehow taken over the Japanese Super Computer? It certainly looked that way.

After Amitch established a download link to our computer, the Japanese Super Computer CPU jumped to 1.95 times its maximum threshold, and started to download the moon hoax evidence it had found. As the computer CPU speed increased, Amitch continued to transmit more and more data. For approximately 27 minutes, the CPU speed increased up to 2.27 times the maximum threshold (limit). Then, all of a sudden, the connection was lost. Over the next several days, we constantly attempted to reestablish a link to the Japanese Computenik Super Computer with no success.

Eventually, however, a connection was made to the super computer, but to this day, no one has been able to reestablish a link to the Amitch System. Without locating the Amitch system, it's impossible to determine if it was indeed the certain programming modifications that possibly caused Amitch to jump to artificial intelligence, or most likely just a programming malfunction. There definitely could have been any number of people working on the Amitch System, and with the data lost, it's impossible to determine which person, or persons, made any program modifications.

However, many people believe it was Mitchell himself that somehow cracked the code, and made the computer reach some kind of artificial intelligence. Such a feat in the twenty first century is extremely far fetched. I can see why the Whizkid Mitchell had something to do with the system crashing. When you ask him if he knows what went wrong with Amitch, he just smiles and says, "I don't know what happened". But nobody believes him. They think he downloaded some type of automatic backup of the Amitch System to his computer and used it to perfect his stock trading system. The strange thing is, shortly after the Amitch system crashed; Mitchell was making a small fortune on the stock market. The stocks he buys seem to somehow magically skyrocket. For example, he bought ERICY and it climbed over 831%. Mitchell bought Perma-fix, and in just a few short months, it went up over 534%. His stocks are still climbing.

Some Of the Amazing Discoveries Amitch Made

Before the Amitch System crashed, its increased computer processing produced exactly the type of evidence we were looking for. Amitch provided so much data information, if put in this book, it would be the size of several large phone books and would take several years to analyze. Therefore, the remaining chapters only contain the most exiting of Amitch's discoveries. For the rest of this book, the remaining evidence will focus on evidence indicating NASA officials and employees clearly knew the moon landings were faked, and took certain measures to conceal the truth from the public.

Some of the evidence Amitch provided in its final moments is truly amazing. An individual can see why humans were never able to figure it out. For instance, one thing Amitch determined was that NASA had many secrets. Furthermore, by specially categorizing systems, they had the entire Apollo moon landing evidence. For all the Apollo mission photography, Amitch downloaded a detailed analysis of the NASA numbering sequence. The analysis shows exactly what methods NASA used to cover up all the evidence that would expose their moon landing hoax.

In these last few chapters, the remaining evidence was provided by the Amitch System, and its exciting to see and easy to understand. However, if you were some kind of super computer, Amitch's theory would make sense to you. Perhaps if you were some person from the future who traveled back in time to study the evidence, you would also understand Amitch's theory. When you think about it, as stated before, the American government would have destroyed most of the data before the year 2026. For this reason, in order to study the moon landing hoax, people from the future would have to travel back in time. In the future, there will be no proof.

In fact, sometimes I feel some of the people working on the Whizkids Project were from the future, and were sent back to the twenty first century to correct this error and help everyone stop destroying the planet for the children of the future. Anyways, if, at first, you don't understand Amitch's analysis of how it came up with the evidence, move on to the upcoming picture. This will help explain the process.

Amitch's Numbering System

The following is a breakdown of NASA's numbering system Amitch determined to be used by NASA for the moon landing photography. Although Amitch's theory on how it came up with this chart may take years for earthlings to figure out, the chart is easy to follow and provides a little more explanation.

Here is Amitch's explanation of how it deciphered the moon landing photography evidence and came up with this chart. According to Amitch, "NASA used a precise numbering pattern that would best equate to a dual multiplex modulated numbering system." This involves a series of smaller sequential numbering patterns, based on primary core sets of numbers coming from one location. A secondary set of primary numbers branches off to create a sub-modulated series of photo tracking numbers.

If you understand what Amitch just said, you might want to call NASA right away. They most likely have a very high paying job just waiting for you. Someone needs to help them answer the overwhelming amount of questions they will be receiving about their moon hoax, and how Amitch came up with the evidence you are about to see in the rest of this chapter.

In simple terms, what Amitch discovered is NASA was using two different sets of tracking numbers for each picture. The photographer taking the picture assigned it a number and sent the photo to the pre-screening department. Amitch program calls this the "Photographer number". The pre-screen editing department sorted the photos for defects and kept the defects photos with them. The pre-screeners then sent the others to NASA's editing department. After the pictures were separated, both the pre-screening department and editing department re-organized the photos before assigning them a tracking number. This second number assigned to each photo is what Amitch called the "Filing number". Below is a breakdown of the various sections of the photographer number and filing numbers.

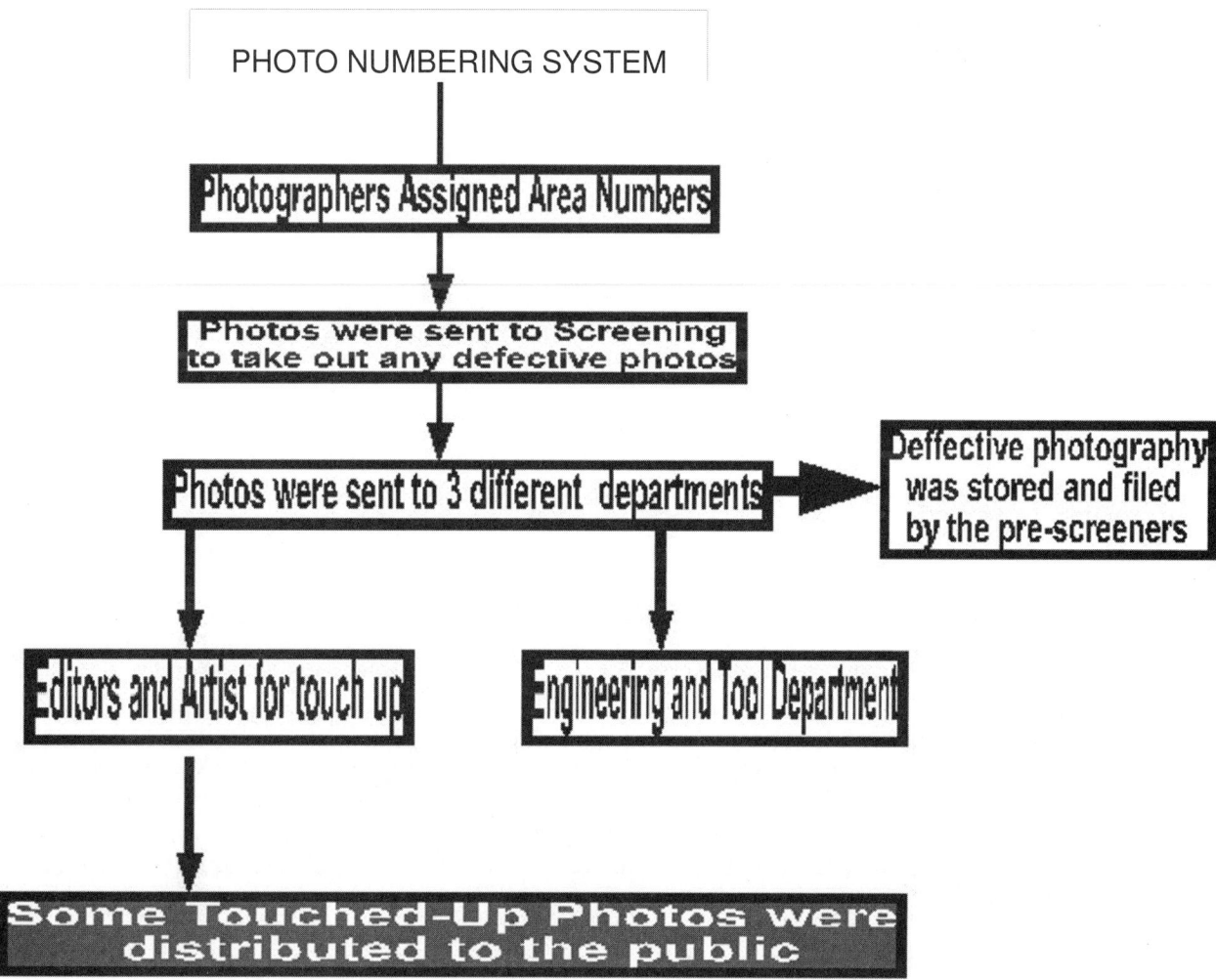

Photographer Number

The Photographer Number looked something like this "ap11-69-KSC-3333", and each section had a specific meaning.

1) The first part of the number contained a mission number such as ap11. This referred to the Apollo 11 mission.

2) The next set of numbers indicated the moon landing simulator the photographer took the photo with, or which simulator the photographer was stationed at. Each moon landing training simulator was given a two-digit number. This number was referred to as an area number. For example, numbers like the following: 69, 70, 71, or 72. Outside the Kennedy Space Center was Area 69, and inside the Kennedy Space Center were several area numbers including areas 70, and 71. For instance, if a photographer working out of the Kennedy Space Center area 69 took a photo at the Taos, Area 19 Nevada, and New Mexico are 54, he used either 70 or 54 for the second number.

For those of you that are wondering if any of the fake moon landing pictures came from the famous Area 51, the answer is no. There were never any photos labeled from the Area 51 location that is related to the Moon landing hoax. They must have been working on other projects like interrogating space aliens and trying to figure out how their space ship technology works and covering up moon hoax leaks.

3) After the area number was assigned to the picture, it was used to determine what portion of the Moon landing the photo was related to. To keep anyone from being able to link the photos together, it appears these numbers were random. The only thing this section had in common was if pictures were taken in sequence, they would generally assign sequential numbers.

4) Finally, if a photo had an extremely obvious defect, the photographers added a descriptive number to the end of the numbering sequence to let the editors know that the picture was really bad. One photographer used a number followed by the three letters "det" for defective, and another photographer used "dmr" for defective material reject (ap11-s70-12533det).

The Filing Number:

The second set of tracking numbers was assigned to each photo by a pre-screen department, or by the editors department. These numbers were 8 digits long ranging between 10000000-30000000. Figuring out how these numbers were assigned, lead to the discovery of a major hole in NASA's numbering system.

Remember, before sending pictures to a central pre-screening department, the photographers developed and assigned each photo a number. The pre-screeners job was to go over all the photos and detect anything that might expose the American's moon landing hoax. The pre-screeners then separated the defective photos and forwarded the good photos to the editing team for artists to touch up.

Based on the scene they were going to use the pictures for, the editing team took the good photos and grouped them together. As an example, photos were put into groups such as the space walk, or the moon rover ride. After the Editors regrouped the photos, they assigned each picture an 8 digit filing number. To eliminate any possibility of someone detecting their numbering sequence, the numbers were randomly separated in blocks of thousands.

So what went wrong? Nobody should have ever been able to figure it out. Unfortunately, NASA did not know that someday some of the smartest minds on the planet would get together and build a special super computer to crack the code to their Moon landing hoax. The Amitch system eventually found a broken link in NASA's filing process. This broken link was related to the defective pictures that the pre-screen department kept. NASA allowed the pre-screening department to use the same 8 digit numbering system as the editing department. Hence, a recognizable pattern was created.

Amitch determined that the pre-screening department made the mistake of separating the photos into various "Defect Groups" based on the type of problem they exhibited. For example, shots were labeled similar to the following: animals in the pictures, stage light showing, and clues left by employees. After that, the pre-screeners made the biggest mistake that opened up the door for cracking the number sequence. NASA assigned each defective group one block of 8 digit numbers and labeled every defective picture in that group in sequential order. Duh, how could NASA have been so brainless? This proved they where looking for the animals in the pictures. It also revealed that they were looking for stage lights. NASA masterminded the entire hoax. No doubt about it.

Listed below are several examples of the numbering system patterns that the NASA pre-screening department used for filing defective groups of moon hoax photos. Clearly, every photo, in each group of numbers, contained the same type of problem, and the next few pages will prove it with matching pictures.

A Few Examples of NASA's Defective Photos Filing System

A) Photo's containing stage lighting problems: 20110000 and 20160000

B) Different items found in mission pictures that would not have been found on the moon: (10010000 - 10010400). These items included wedding rings, plastic beer cups, and cigarette butts.

C) The Animal Kingdom: 20149660-2049670. This was a series of photos containing earth animals in the alleged NASA moon pictures.

D) Un-edited photos of stage prop tools: judy01-judy200. These had many earth training scenes. They matched moon scenes identically.

E) Moon Rock Photo's with different types of discrepancies: 20147700-20147720

F) Photos with Secret Messages written in the moon surface as clues, in hope that we would find them as15-82-11050-as15-82-11190. Work include FAKE, At Kennedy Space center, Artist Signatures embedded into the painting and much more. These clues are not easy to detect in printed material and for better clarity you must view these photos on NASA's web site, www.whizkids.tv or www.moonbloopers.com

G) Problem item found on moon rover including; a snake, cat, and desert tumbleweed hung up on back of rover while on moon. What was NASA thinking? as15-82-11196-as15-82-111203. For better photo clarity you must view these on NASA's web site, www.whizkids.tv or www.moonbloopers.com

H) Sample of Astronauts (actors) wristband movie script booklets: Impcuf # 01 -18.

I) Stage Lighting adjusted too high. Thus, showing improper illumination of the moon surface and backdrop: 20128960-20129000.

J) Series of pictures that show how scenes were transformed by NASA artists into new ones: 20153740-20153800.

Back then, NASA decision makers had no crystal ball and could not have known what the future technologies would bring. That's why, they decided to scatter the defective photos around to different government agencies, museums, and other countries. They believed no one would ever, in a million years, be able to travel to all those places, and see all the defective pictures. Besides, if someone did travel to all these places, there was no way they would be able to figure out the defective numbering sequence. Since these places were not permitted to display the photo ID numbers, they were to stay in the museums private records.

Back in the 1970's, how could anyone know that the Internet would come, and someday all these different places that received the defective photos would make them available for everyone to see? Many years later, unaware of any defective moon landing photos being scattered throughout the world, NASA implemented a new Internet policy: All NASA photos, displayed on the Internet, were required to contain their assigned photo ID number. For goodness sakes, no one back then would have even considered the photos with ID numbers a threat. In addition, they could not have imagined that someday a computer system would be capable of deciphering the numbering sequence. Back in the 1960's and 1970's, the basic electronic calculation was in its infancy.

Chapter 15- Part 2

Examples showing NASA used a coding system to hide evidence about their Moon Landing Hoax

Unfortunately for NASA, today, these defective pictures are available to the whole world via the Internet. Within a few hours, anyone can collect many of these defective photos. It's now even easier than that. All someone has to do is read the remaining pages of this chapter. By doing so, they will observe countless examples of how NASA faked the moon landing photography and attempted to cover it up.

Improper Stage Lighting Conditions: 20110000-20160000

While attempting to produce that perfect looking moon landing picture, NASA photographers frequently experience problems getting the lighting correct.

Unless someone purposely filed them in that order, how could every photo between 20110000-20160000 exhibit the same problem? Every image has an over lighting condition. There is no way it could be a coincidence that all the photos in a series of exactly 50,000 numbers could all have the same exact problem. These moon landing photos had to have been assigned their individual numbers because of the problem discovered in the pictures. This reconfirms Amitch's theory that NASA pre-screeners separated the defective pictures into individual groups based on the various types of problems. Next, they assigned each defective group a block of numbers. Finally, the pre-screeners assigned each defective picture a number from that block of numbers.

Below are a few samples of the more obvious pictures that were assigned to this grouping because they exhibited an excessive stage lighting problem.

Deciphering NASA's Coded System For Hiding Their Incriminating Evidence

Stage Lighting Problems- 20125000 and 20130000

Amitch determined that NASA used the numbers between 20125000 and 20130000 to store photos that exhibited over-lighting conditions that occurred when attempting to illuminate certain objects, such as the American Flag, the LM spacecraft, or the astronauts.

There is no way this could be considered a coincidence. How could every picture assigned in this NASA series of 5,000 numbers contain an over-lighting condition in the direction of these objects? In an attempt to illuminate these objects, it could only have been a spotlight. Below are a few shots of the more obvious pictures that were assigned to this grouping because they exhibited this particular type of lighting discrepancy.

Another group of improper stage lighting conditions: 20120000-20160000.

From the very powerful stage lights, there were several different types of lighting problems produced in the moon landing photography. To categorize the portraits into specific groups, based on the type of lighting problems, NASA editors used extra efforts.

After discovering all the photos in one group were from random missions, it was possible to determine how NASA editors assigned each of the defective pictures their own tracking numbers.

* First, based on the various types of problems, the pre-screeners separated the defective pictures into individual groups .
* Secondly, they assigned each defective group a block of numbers.
*Thirdly, they gave each defective picture a number from that block of numbers.

F flag bright -astronaut dark

10135264

20130616

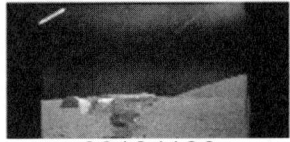
20134130

20134616

20134628

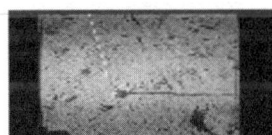
20134640

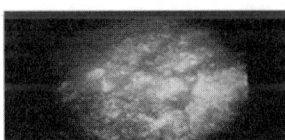
20135152

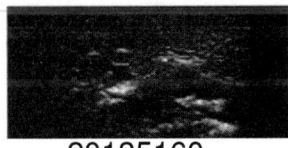
20135160

20135220

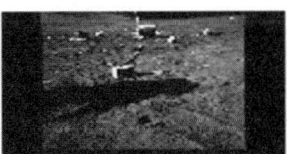
20135229

201345232

20135237

20135240

20135254

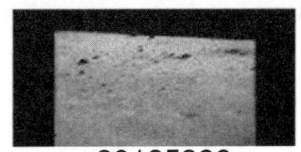
20135399

20135611

Improper Moon Surface And Backdrop Illumination - Continued

Discovery of overhead stage light being used in Moon Landing photography

After scanning through the thousands of photos between 2012000-2016000 that contained excessive lighting conditions, the actual stage light can be seen in several of the following images. Just as one would expect the stage lights were positioned above and angled to the side of the camera. It appears the film crew wrapped lights in some kind of black tarpaper with black tape.

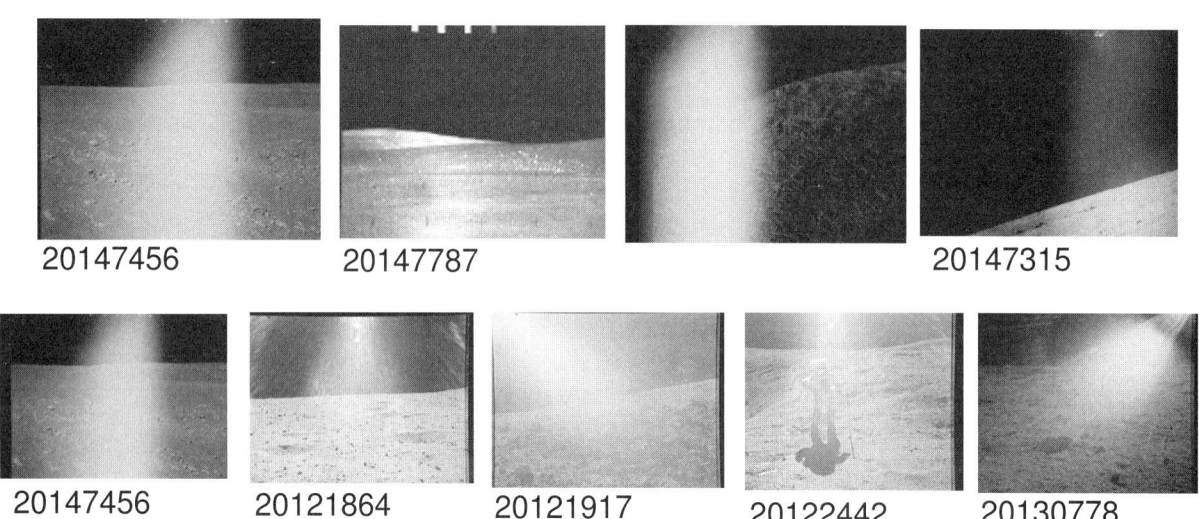

Stage Lights Hidden In Rocks

One of the most exciting stage lights discovered was the moon landing photo # 20122176. It revealed a stage light camouflaged between moon rocks. Below are a few samples.

Deciphering NASA's Coded System For Hiding Their Incriminating Evidence

NASA had several divisions working on the construction of the Moon landing hoax and each had different tasks. One of them was in charge of developing stage prop tools that would be used by the astronauts (actors) on the artificial moon surface.

The tool department labeled their photos with the prefix "j.u.d.y.". The thing that makes these photos so special is not that we can see all the fancy tools they came up with, but they were never sent to the editing department to be touched up. For this reason, they exposed the moon hoax.

Here, some of the "j.u.d.y." images contain some of the most powerful evidence proving the moon landings were faked. With a glance, one can see some of these pictures are missing crosshairs and have other discrepancies.

The Tool Department's Photos

NASA ID# judy39

NASA ID# judy48

NASA ID# judy60

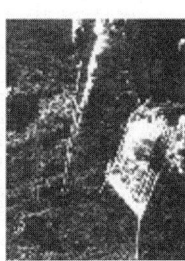
NASA ID# judy62

NASA ID# judy61

NASA ID# judy67

NASA ID# judy68

NASA ID# judy69

NASA ID# judy73

NASA ID# judy77

NASA ID# judy85

NASA ID# judy91

Evidence suggests NASA's tool department's photos were not originally intended to be distributed to the public, so there was no need to have artists touch them up. Unfortunately for NASA, over time, the people that knew the moon landing pictures were defective retired. As a result, new employees, unaware that the moon landings were faked, later made the pictures available to the public.

Since these tool department photos were unedited, they revealed some of the most hidden secrets of how the moon hoax was masterminded. Many of the j.u.d.y tool photos matched the graphics taken here on earth.

15-13

The Moon's Animal Kingdom

At first it seemed the Amitch System had malfunctioned when it suggested NASA editors had even established a numbers system for photos containing earth animals discovered in the photos. Although now we can see many of the alleged Moon landing picture contain animals that managed to find there way onto the outdoor movie studios. In many cases it would have been impossible to detect any of these animals without Amitch detailing exactly where and what to look for. Even then in many cases it required examining the photos with a powerful magnified glass to confirmation Amitch's findings.

Below is one of the series of numbers containing earth creatures 20149660-2049670. There are so many moon landing pictures with Earth's creatures in them, a whole chapter in this book has been dedicated cover the animals in more details. The chapter is called "THE ANIMAL KINGDOM". However, below is one of the series of numbers containing Earth creatures 20149660-2049670 Amitch discovered as an example.

Jack Rabbit Resting on moon

20149666

Lizard batthing on rock maybe

20149667
Gold Tegu

Snake

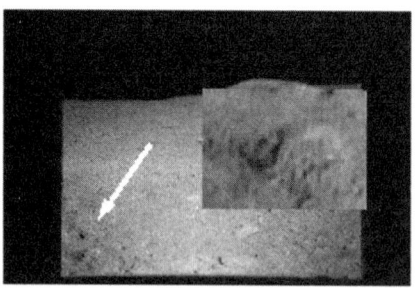

20147668

Dog finds Lizard on Movie Stage   "Lizard vs Dog" take one

20147669

Objects Found In Pictures That Would Not Have Been Found On the Moon (10010000 - 10010400)

Considering the fact that numerous people were working on the movie set, it would not be uncommon for someone to accidentally leave an object on the stage. One of the most important jobs of the photo pre-screeners was to search for items left on the artificial movie stage. This was to prevent people from later finding and exposing the moon landing hoax. Pre-screeners looked for items such as the message written on the moon surface, and the word faked added to the astronaut's backpack. Remember, the Australians claimed to see a coke bottle roll on to the live Apollo moon landing film, which many believe was latter edited out by NASA.

The list goes on and on. As you'll see, it appears NASA had a problem with many employees intentionally trying to leave clues about the hoax. NASA employees were leaving coins, pictures, notes, kid toys, and even wedding rings in the sand as clues. The important thing to remember here is that we can prove NASA knew people were attempting to blow the whistle. Why did the editors group the photos containing foreign objects together? Below are

Seashell 20135252

Mixture of different Earth trash items

10100250

10100256

10100257

Unkown Items

10100037

10100062

10100056

10100071

Cigarette Bud ?

10100138

10100142

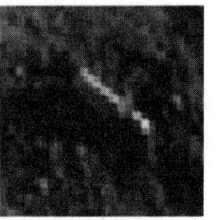

Deciphering NASA's Coded System For Hiding Their Incriminating Evidence

Below are a few highlights of some other miscellaneous items found on the moon surface. They don't fit into this series of numbers; however, they should be referenced at this time to show how easy it was for objects to go undetected by NASA's editing team.

Two Human Fingers:

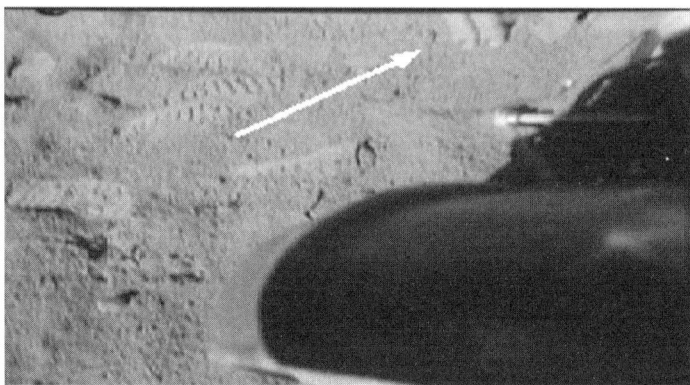

NASA ID# 10010

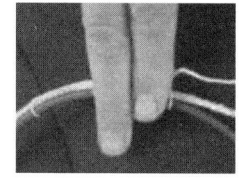

Two Fingers

Movie Producers Chair Camouflaged in Black:

Producers Chair

NASA ID # as17-151-23188

Message Clue: A note left by astronauts, which appeared to be a clue that the moon landings were faked.

NASA ID# 10074365

NASA ID # 10074363

15-16

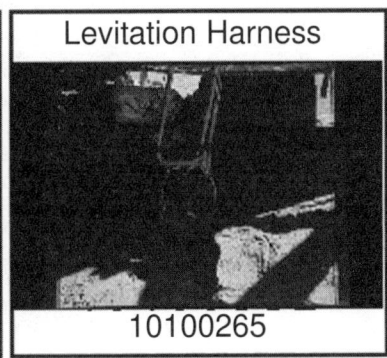
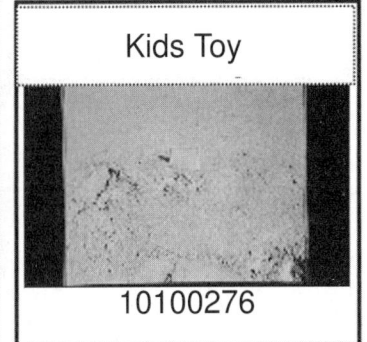

Stage Crew Footprints

Objects In Front Of Moon Pictures

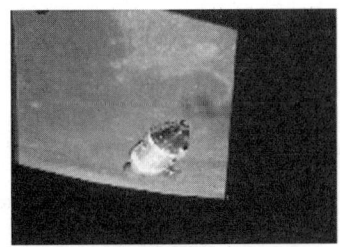
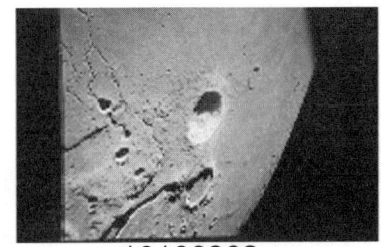
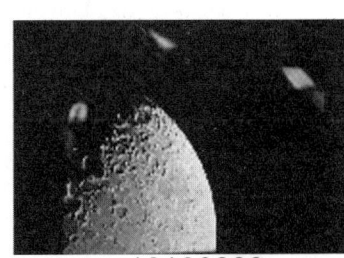

Deciphering NASA's Coded System For Hiding Their Incriminating Evidence

Messages Written in the Moon Dust

Every NASA photo between as15-82-11110-as15-82-11160 contains some type of secret hand written message left by NASA employees in hopes that some day their clue would be found.

Many clues were coded messages written in the Moon dust and are very difficult to decipher their meaning. Yet some are very obvious and easy to understand. Some are a combination of letters others are the words Fake, Hoax, Not Real and some are signatures of the artists. Others are a combination of short sentences such as Kennedy Space center, on Earth, A Dog, and much more. Once again, they are difficult to see in printed material and for better clarity you must view these photos on NASA's web site or www.whizkids.tv or www.moonbloopers.com

**Word "FAKE" written on side of Rock
NASA ID # s15-82-11138**

Code message we killed KENNEDY ....

Message reads "We killed Kennedy...."
NASA photo ID # NASA ID # as15-82-11160

**Many Word Painted into this Photo
NASA ID# as15-82-11135**

**Several writings in this photo
as15-82-11140**

Movie Script Attached To Astronaut's Wristbands

There are many pictures that show the astronauts using what appears to be actor movie script books and pre-made maps of the alleged mission landing sites. The subject of the content was always contained in these script books. Although no official NASA explanation as to what these movie script books contained were ever found. Amitch deciphered a number sequence for the pages in the script books. It revealed the entire Moon landings were nothing more than a pre-made movie. Below is the official Apollo 12 book with the sections considered unacceptable for readers covered up. This is in no way a joke.

 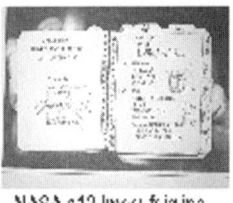 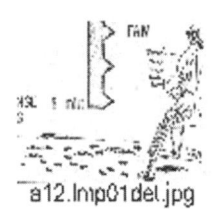 

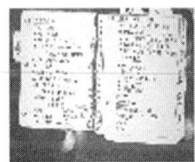  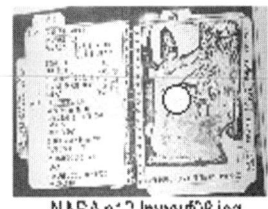 

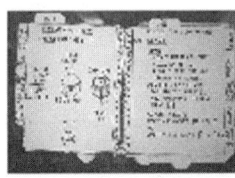   

 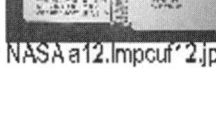  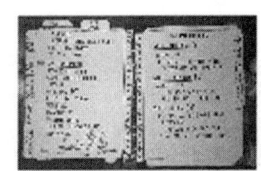

 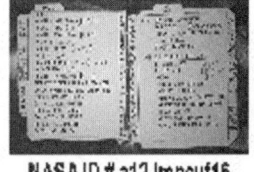 

Could There Have Been A Wedding On the Moon?

After searching thousands of photos, a problem with Amitch's artificial intelligence was discovered. Through a series of data checks and balances, Amitch somehow determined that a wedding took place on the moon. From this, Amitch provided this combination of Moon landing mission illustrations as evidence. As you can see, Amitch found a wedding ring, groom, bride, red bible, and the receptions party supplies.

Regardless of how convincing the evidence may seem, there was never a wedding on the moon. Nevertheless, the visual renderings demonstrate the limited capabilities of earth's artificial intelligence technology. Unlike humans, computers lack common sense and base their conclusions strictly on raw data.

One can easily see why the American's nuclear weapon silos required the final launch code to be provided by a high ranking government official. World leaders make extremely difficult decisions, such as blowing up the planet. Naturally, a computer analyzes all possible out comes. Because of this, it would determined not to launch nuclear bombs in defense. Believing, regardless of which side wins, if half the world's population survived, this would be better than none.

CHAPTER 16

Methods Used To Fabricate the Moon Landing Scenery

Chapter 16
Methods Used To Fabricate the Moon Landing Scenery

This chapter will highlight some of the most serious types of problems discovered with NASA's alleged moon landing photography taken on the moon surface. What you're about to see related to the NASA alleged moon surface will prove conclusively the Apollo astronauts never stepped foot on the moon. The evidence in this chapter shows how the artificial moon surface was constructed in great detail; including the various clues that led to this discovery.

These next photos reveal how, prior to the astronauts (actors) arrival, the stage crew set up the artificial moon landing scenes. These pictures, supposedly taken on the moon, have something very wrong with them. They have missing tire tracks leading to, or from, the Rover vehicle.". The lack of tire tracks is a big clue that helps prove the moon landings were faked. This indicates the moondust soil is some case was added after the Rover Vehicles were placed in their locations.

Also, we know the moondust mixture was most likely added shortly before the astronauts arrived on the movie set because the astronauts were leaving bold footprints elsewhere on the scene. Because of this, when filmed, the soil was still moist.

Rover Pictures Missing Tire Tracks

Apollo 17 as17-146-22367

Apollo 17 - as17-143-21933

The remaining question that needs to be answered is could the astronauts simply have picked up the Rover vehicle and carried it to a new location for one of many reasons. The answer is yes, the moon rover side rails have built-in chassis handles to the right of the wheel. The handles are roughly halfway between the front and back wheels on each side of the Rover. The handles were level with the balance point of the Rover.

The handles were provided to allow the astronauts to pick the Rover up and turn it around. The Apollo Lunar Surface Journal discusses this in the Apollo 16 commentary, where Eric Jones discusses it with Charlie Duke. However there are other clues the help prove this is a movie studio and the rover was carried to it's location by the astronauts.

For example take the Apollo 16 photo on the left, it is not only missing the tire tracks it is also missing the astronauts' footprints. While the picture on the right shows the astronaut driving the rover in dust and producing no tire tracks with dust falling off the tires.

Moon landing pictures were missing tire tracks and some footprints.

No tire tracks while driving.

No tire tracks or footprints.

Apollo 16- NASA ID# 20134558

Apollo 15 - as15-85-11471

The picture on the left was taken prior to the astronauts filming in this particular movie scene. While the image on the right is another example of NASA photographers using a "Cut and Paste" composite picture technique to create the portrait.

A Close Examination Of the Moon Surface Also Revealed Moon Landings Were Faked

There are without a doubt thousands of obvious clues, clearly visible to the naked eye, which proves NASA's photos are fake. The discovery of the letter "C" in the Moon landing photography is one perfect example. The use of the letter "C" on film props is a very common studio identification practice. It is used to mark where the center of a stage scene should be, which provides photographers their reference point for filming. The famous C-Rock pictures are shown below.

This image from the Apollo 16 mission has a rock positioned in the foreground with a very obvious letter "C" on it. If our theory is correct that NASA filmed the Moon landings in their Apollo training simulators, then it would not be un-expected to see something like this.

Letter "C" on Rock
NASA original photo ID # j.u.d.y.69

Here is the exciting thing about this "C" rock story. After it was discovered on this rock and brought to NASA's attention, they quickly switched it with an edited version, this time with the "C" marking removed. This shows that some NASA officials are well aware of the Moon landing hoax, and are continuing to cover up evidence each time something is discovered. How could they continue to do this? NASA would need a whole department of people dedicated to taking every action necessary. As the question of proof is becoming more and more complex, with the advancement of scientific technology, one would think this secret government group would be enormous by now. Now that modern science is proving over and over again that the alleged Moon landings were simply impossible, this group must also be working extremely hard to continue devising new methods of distorting scientific advancements in order to keep the hoax a secret. But does NASA even have a group this large with such capabilities. Yes, you guessed, at AREA 51!

Below are the two NASA Apollo 16 photos in question. One shows the rock with the letter "C" placed into the rock and the other does not. Clearly one is the original and one is not, but which one is it? Did NASA's Moon landing watchdogs at Area 51 edit the "C" out of the original picture?

"C" on rock

Letter "C" on Rock
NASA original photo ID # j.u.d.y.69
(also ID# as16-107-17445)

NO "C" on rock

NO Letter "C" on Rock
NASA original photo ID# as16-107-17446

According to NASA's Moon Landing photo numbering formula, determined by our hyperthinking super computer system. The rock containing the "C" is the original Moon landing photo. Both these images clearly follow a well determined path of NASA's tracking numbers assigned to all the moon landing photos.

First, the photographer that took the picture assigned his location number j.u.d.y.69. This number refers to photos taken outside the Kennedy Space Center. We also know that the original purpose of this photo was to get a good picture of the tool rack on the back of the rover vehicle. How do we know this? Simply put, the camera was focused on the tool rack, and the tracking number assigned to this photo "j.u.d.y.69" was a NASA tool department number.

Since the j.u.d.y.69 photo originally belonged to NASA's tool department, whom we know never edited their photos because they didn't intend to be distributing any of them to the public, this is the original photo. Remember, that NASA's tool department j.u.d.y. photos were released accidentally many years later.

To give NASA the benefit of the doubt, lets imagine they are telling the truth and the Letter "C" was never on the rock in the original Rock photo.

NASA claims that the "C" was just a piece of hair or thread from someone's clothing that accidentally landed on the original photo during the copying process. If NASA was correct then this would be the first piece of evidence suggesting anything in this book is incorrect. Now we can't have that can we? Everyone has put too much hard work into this project to let NASA off the hook with such an idiotic response. Besides, this also would suggest our super computers "Amitch" system, which was designed to help decipher the NASA Moon landing hoax was not perfect. But how can anyone really know who is telling the truth about which photo is really the original? Is it the honorable NASA establishment or modern day science and Whiz Kid geniuses, who is right?

One way to make a final determination if NASA's theory of the hair on the picture claim is correct, is by seeing if there are other NASA Moon landing pictures that contain the same Letter "C" stage marking symbol. If it is not present in any other NASA Moon photos then we will declare NASA the winner. If many other photos contain the same strategically placed letter "C" in them, then we will declare the hoax theorists the winners. And the winner is the Whiz Kid geniuses.

Based on the overwhelming amount of additional Moon photos containing the same identical letter "C" movie stage markers, NASA has some serious explaining to do! The following pictures are only a few examples of Moon landing photos containing the "C" movie stage marker

Samples of Additional Moon Landing Photos Containing the Letter "C".

Movie Stage Markings
NASA ID# as15-82-11135

This photo contains other markings in addition to just the "C". The bigger words read: TV 1XXM CL, which would indicate the proper camera LUX magnification adjustment at the center location of the stage.

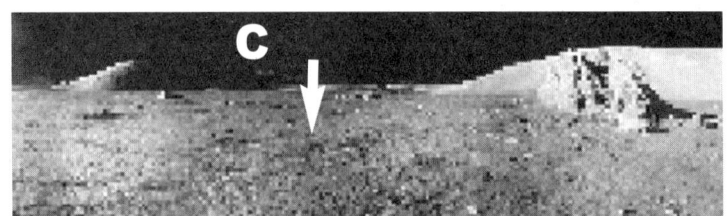 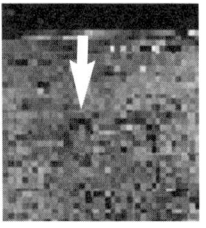

Center Stage "C" Marking was placed into the center of Moon landing site scene. Indicating a special C-ring was placed on the scene before the final layer of Moon dust was sprayed on before filming. This would be very visible to the camera crew during filming and easy to cover up during final editing

Although the letter "C" is present throughout a great deal of the Moon landing photography, there are still going to be a few people that will find it hard to believe NASA has been editing the errors found in their Moon landing photography all these years, however this next piece of evidence should clear things up for any of those remaining die hard Moon landing believers.

There is one way to prove for sure that NASA's photo without the "C" is not really from the moon. All we have to do is prove both the pictures are fakes. And that should not be too difficult, since there is something much bigger that's wrong with both these pictures. When NASA leaders see what is really wrong with these pictures, they will probably faint, it is almost unimaginable. This problem is so widely spread in NASA's Moon landing photos, an entire chapter of this book has been dedicated to this very significant problem. There you can learn exactly why these photos are fake and what is really wrong with them.

Okay, if you insist, here is a sneak preview of what is going to be covered later in the book under the chapter "Whistle Blowers".

First of all, the hoax conspirators are right, the "C" was on the original photo and was never intended to be released to the public. Also the letter "C" does indicate the center stage marker and is a legitimate error in the photo. After the photo was released and the letter "C" was discovered, NASA needed to fix this photo quickly and get a corrected version back out to the public ASAP. If they did not, they would have attracted further speculation of the legitimacy of their Apollo Moon landing photos.

Under pressure to remove the "C" and get the edited photo back out to the public, then NASA's watchdogs working at the time, screwed up. They assumed the "C" was the problem with the photo, because it is so obvious. What NASA's editors forgot to do was cross-reference the second identification number assigned to this photo. If they had, they would have realized the error number belongs to a special group of photos, with a very unique type of problem. This type of problem is so serious, if any of these photos were to reach the public and the problem discovered, certainly NASA would eventually be forced to confess to their moon-landing hoax.

The "C" rock photo like so many others belongs to the special group of classified photos called the "Whistle Blowers". The word "Whistle Blowers" refers to NASA's employees that worked on the Apollo Moon landing movie, who left clues. Hoping that someday people of the future would find their clues and expose NASA's Moon landing hoax. Well congratulations Mary, who left this clue for everyone to find, you did it.

If you look very closely at this Moon landing image with a magnified glass, you'll see that Mary wrote a special message in the moon dust with a sharp object, most likely a pen or pencil. The message is a little distorted however it appears to read, "LOOK mommy, love Mary".

Can you see the message? Probably not without a magnified glass, it's located between the back wheels of the moon rover.

This is just one of many examples just like this where people working for NASA at the time wrote secret code messages into the moon dust in hope someday they would be found. Originally NASA watchdogs caught most of these. The funny thing is, over the years, NASA has been gradually forgetting what the moon-landing photo numbers stand for and have started distributing them in error. In some way one has to feel sorry for NASA, billions upon billions of tax dollars wasted and they still can't get it right.

The 'N' Rock

If the "C" rock was the only example of an American letter found on the moon, NASA could dismiss it as a hair, or fiber that somehow got on the paper when a copy of the negative was made.

However, several more of the Apollo Moon landing photos show rocks with other markings from the English language. For instance, another excellent example of unusual markings on the moon rocks would be the "N" Rock. This provides additional proof that the moon landings were filmed on a movie stage here on earth.

The "N" Rock
NASA ID# 20147705

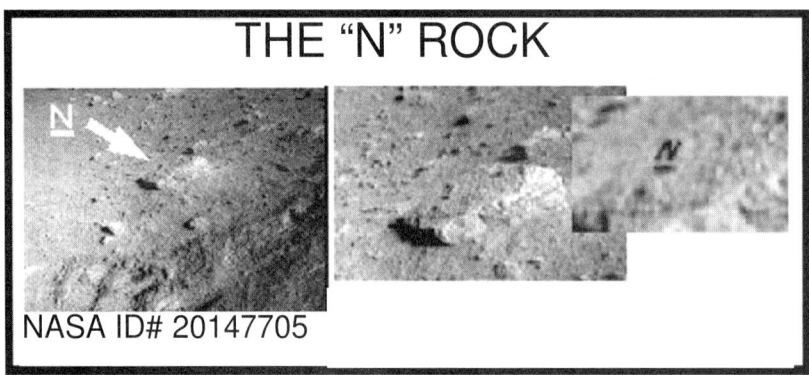

It is virtually impossible that this is a mere coincidence. Two different rocks, in two different locations, marked with two different letters from the English language on the Moon. The most logical explanation is the "N" on this rock represents a movie stage marker. This indicates the rock was placed on the North end of the artificial Moon surface under the Langley lunar simulator crane. This theory can be confirmed by the "S" found in the next picture of the opposite end of the Moon Landing stage, which would be indicating the South end.

Or maybe space aliens who travel frequently between the Earth and Moon are now using English as a second language. Clearly, these markings suggest some sort of stage prop marking system. Under these circumstances, any other suggestion would simply be ridiculous. Many of NASA alleged moon landing photos contain the exact same type of stage marking symbols in them.

Methods Used To Fabricate the Moon Landing Scenery

This next group of pictures reveal that the "N" was used as a stage marker on other rocks and moon surface scenes. If you need more evidence to be totally convinced that these are indeed stage markers, you can view the other similar photos at NASA's website or at www.whizkids.tv or www.moonbloopers.com

"N" Stage Marking Found on More Rocks

N rock: AS15-82-11190

N rock: AS15-82-11189

"N" Stage Marking Found on Moon Surface

"N" Markings AS15-82-11129

"N" Markings AS15-82-11129

"N" Stage Marker AS15-82-11216

"TEN" Markings AS15-82-11131

While searching for evidence of NASA filmmakers using these letters as movie stage markers, it was discovered that other people working on the studio site were writing things in the Moon dust. There were hundreds of words written on the surface, the most popular one was FAKE! Coded messages and Autographs were also very popular.

These next Apollo Moon photos show the word FAKE written all over them. That was not meant as a figure of speech, these Moon landing photographs really have the word fake written all over them.

The NASA Moon Landing Pictures Below Have the Word FAKE Written in the Moon Dust

Word Fake written on side of Rock
NASA ID # AS15-82-11138

a16-107-17500

Fake Written as Clue: AS15-82-11117

FAKE written all over it. AS15-82-11-11169

Chapter 16 Part 2

How NASA's Moon Landing Movie Stages Functioned

Moon Rocks Are Made Of Plaster

After realizing the moon landing missions were staged here on earth, the remaining unexplainable moon anomalies could be answered using some basic common sense.

Rather than trying to determine how the overwhelming number of strange conditions seen in the moon photography could have possibly taken place, I decided to change my approach to analyze how these conditions could have occurred on a movie stage. From that prospective, everything seems to make perfect sense. The problems discovered with the Moon landing photos can be attributed to common difficulties while filming a movie at these NASA training simulator locations.

White Rocks:

For instance, this explains why the astronauts found white spots exposed on many of the moon rocks. Here is the problem with the NASA pictures shown below. The discovery of the natural white color of the rocks would be unlikely since almost everything is covered with compacted moon dust and meteorite particles. NASA's astronauts claim of finding these amazing white rocks on the moon is too ridiculous to even consider.

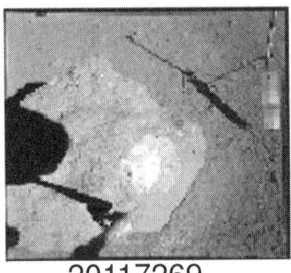
20117269

20116952

as16-116-18653

20116729

There is solid evidence suggesting that, when the astronauts accidentally touched the rocks, they caused the artificial moon dust to come loose, which made the white spots. Because of this, the astronauts uncovered the plaster stage prop rocks. This appears to have been one of the biggest problems for the NASA photographers. Apparently, the moon dust was a fine mixture of moistened sand and volcano charcoal ash that, prior to the astronauts filming each scene, was sprayed onto the site. Under the stage lights, the dust dried quickly and came loose from the rocks very easily. This was a huge problem. If you have ever seen a machine that blows insulation into a house, that would be a good example of how the moon dust sprayer would have worked.

As the pictures above show, the actual final layer of moon dust that was spread over the stage platform was very thin. If the astronauts were not careful, they would rub off the dust and reveal the rocks were made of white plaster.

The picture to the right not only shows the artificial moon dust wiped off the rock, it has also been autographed by someone working for NASA. This would explain a lot of the Apollo Astronaut's comments describing the artificial moon surface. For instance, astronaut Neil Armstrong stated that the moon looked like plaster paris. It appears Astronaut Neil Armstrong, when he made this next statement while supposedly on the moon, was leaving us clues. He declared "The surface is fine and powdery." "I can pick it up loosely with my toe." This was a common American saying. Basically, it meant that to darken the soil mixture, a sanding surface and charcoal ash was added. "It does adhere in fine layers like powdered charcoal to the sole and sides of my boots." Powered charcoal sticking to his boots indicated soil was moistened. "I only go in a small fraction of an inch. Maybe an eighth of an inch." This statement confirms the thin layer of sprayed moon dust observed in many of the pictures. "But I can see the footprints of my boots and the treads in the fine sandy particles." This announcement reinforces a moistened surface. If the surface was not moistened, a footprint would not be possible on the moon. "There seems to be no difficulty in moving around as we suspected. It's even perhaps easier than the simulations." From this speech, Armstrong was trying to give us clues. In addition, he was informing us that the astronauts were actually at the NASA training simulators.

Artist Autograph NASA ID # as15-82-11130

Astronaut Armstrong goes on to describe the surface in more detail, "It's like much of the high desert of the United States." Armstrong is giving us a clue where the filming took place. This just happened to be where the moon background scenery came from.

All signs indicated this was an attempt by the Apollo Astronauts to leave clues for us to find. To be sure, you would have to ask them that question. If they're not willing to fess up to the hoax, don't worry. It's not a big deal. This next photo will do it for them.

Not only does this next moon landing images show the astronauts scraping off the artificial moon dust, it also reveals the white plaster. A NASA editor even notes that, by touching the rocks, one of the astronauts contaminated the stage scene. This caused several noticeable grooves in the moon dust coating, which exposed the white plaster of the rock. One can only assume that NASA published this photo because after 30 some years, no one knew what the markings meant.

a17-21496lbl

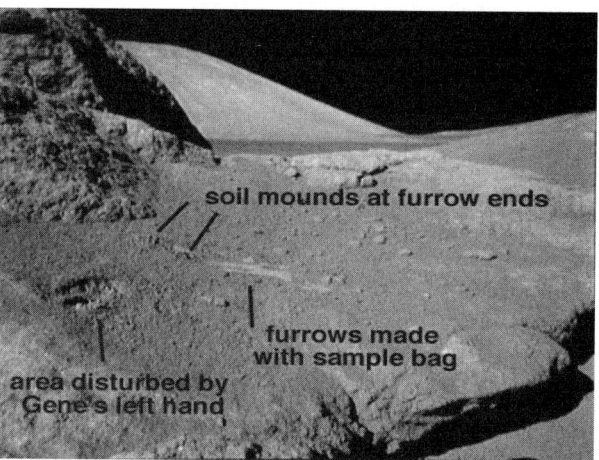

a17-21482lbl

Another Dog Spotted on the MOON

By far, the most convincing evidence the white spots on the rocks are a result of the artificial moon dust being scraped off stage props is in the next picture. While supposedly on the moon surface, one of the Apollo Astronauts is attacked by a black dog. Look at the two pictures below on the right.

ID# 10075624

ID# 20147713

ID # 20147714

Notice how, in front of the dog, in the middle picture, the rock is completely covered in gray moon dust. However, when the dog jumped over the rock and took the astronauts stick away, it wiped a large section of the moon dust covering the rock and revealed it was only a white plaster stage prop rock. One would have to assume a NASA editor mistook the dog for a shadow and accidentally released these photos .

Bold Footprints Are Scientifically Impossible On the Moon

What about all the astronaut's footprints seen on the moon surface? They don't stick to the general accepted principle that the moon's surface was very dry and dusty. Lets think about it for just a minute. How could footprints, so detailed, be seen on the moon surface without any humidity or moisture in the soil? By themselves, these footprint findings prove that the moon landings were faked here on earth.

The next set of portraits are alleged to be Neil Armstrong's first few steps on the moon.

To make such perfect looking footprints, it would take a lot of water. Without a doubt, the moon contained no water in the liquid form. The lunar surface is predominately composed of materials that fall under the general category of silicates. On a very small molecular level, silica does have a slight natural tendency to bond with other silica. Nonetheless, to bond something even as small as a fingerprint, would be an amazing feat in outer space. One of the few substances that would leave footprints so bold as these would be a damp sanding and ash mixture, which all evidence so far indicates was used.

 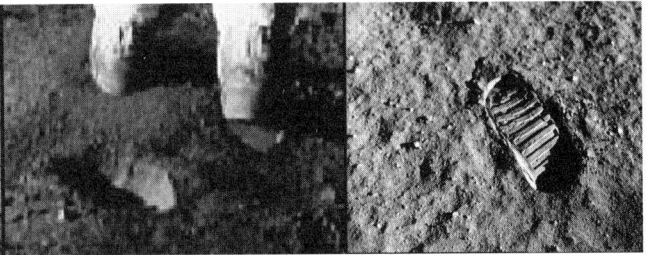

Close-up of an astronaut's footprint in lunar siol. Photographed with 70mm linar surface camera during the alleged the Apollo 11 extravehicular activity on the moon

How could NASA expect anyone in the scientific community to believe silica molecular bonding could create an impression of the identical object such as an astronaut's footprint, and then hold its shape indefinitely?

Could these footprints have been the result of some unexplainable rare soil spot on the moon surface? That's simply not possible either. In almost every photo taken of the astronauts on the moon, there were many examples of recognizable footprints.

Speaking of footprints, who's are these in the picture of astronaut John Young from the alleged Apollo 16 mission? The narrowed footprints have been made by shoes other than an astronauts boots. The footprints either belong to the photographer, a stage crew person, or Mr. Young had an unexpected guest for dinner on the moon.

The footprints look like they matched these two guys seen standing behind the astronauts at the Kennedy Space Center training simulator. Before the astronauts supposedly went to the moon, this picture was taken.

Stage Crew Footprints Found on Moon

  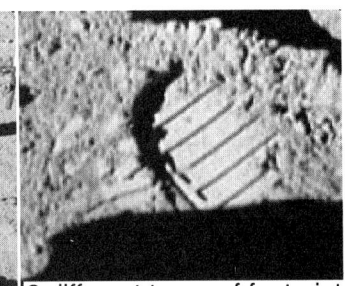

2 different types of footprints

ap16-72-h-100   NASA ID# 20121894

More Footprints Mysteriously Disappear

Another strange thing about the alleged moon landing photos was discovered by Whizkid Tony (age 11). Tony made a startling discovery that opened a whole new group of evidence proving the moon landings were faked. Look closely at the next set of snapshots, and you'll notice there are NO footprints leading to, or from the MOON TOOLS the astronauts are photographing. If there are no footprints, then how did these tools get there?

20134557   20133932   20124191

Apollo 16 - 20134558

as16.110.17960

Thanks to Whizkid Tony's observation, once again, there is proof that the moon photography is fake. When compared to previous photos, there is something very different about these pictures. There is no evidence that the Astronauts were pasted into these pictures; they are really at this location. Tony discovered rare still pictures taken during the live action filming of the astronauts under the Langley Lunar Simulator Crane. It appears the foreground is some type of artificial moon soil, and the background is a painting.

This theory can be confirmed by the fact that the moon rover vehicle is sitting in the middle of the stage without any tire tracks leading to or from it. There is no way for the Rover to be dropped from the sky, or carried to each spot. How can there be no tire tracks leading to or from the vehicle?

Under close examination, with a magnifying glass, one can see there is soil built up around the tire. This is a condition simply not possible on the moon without tire tracks. These circumstances fit perfectly with the theory that these pictures were filmed on earth under the Langley Lunar Simulator's Crane. Bear in mind, how the moon rocks showed that a final layer of artificial moon dust was spread onto the moon stage. The soil built up around the rover tires, and the lack of tire tracks from the rover, indicate the moon dust would have been added the same way.

No Tire Tracks

NASA ID#as17-134-20443

Before the thin layer of moon dust theory can be considered to be a fact, there is still one question that needs to be answered . If the moon rover vehicle was being used as a pre-positioned stage prop, then it would had to have been set to the customer "free-wheel" position. This "free wheel" feature would result in the de-coupling of both the automatic and manual braking systems. Hence, the systems would have been non-functional. For "free-wheel" de-coupling mechanical operations, refer to the NASA Lunar Rover Vehicle Operation Handbook pages 5 through 12.

Since the rover vehicle was often placed on small man made hills on the stage platform, what was to stop it from constantly rolling down these hills with the parking brake system disconnected? Doubtless, just as the astronauts did when they touched the top layer of moon dust, the rover vehicle would have contaminated the newly sprayed layer of artificial moon dust. That surely would have been a major problem for the movie producers since the Moon Rover vehicles were not equipped with parking brakes. Also, there is nothing in front of the tires to keep the Rover Vehicle from rolling away.

This was one of the first problems the Amitch System was asked to solve during the early stages of its development. Amitch was asked: since the rover vehicle braking system would have been decoupled from the traction drive, what prevents the rover vehicle from constantly rolling down the hills on the movie stage platform under the Langley Simulator?

The Amitch computer program went right to work; it looked for an answer. After a few minutes, it suggested that a wire attached to the girders (crane structure) was keeping the rover vehicle from rolling down the hill. Amitch went on searching the data bank of Apollo moon landing evidence and eventually dismissed that theory. Next, Amitch suggested that forced air from strategically placed fans might have kept the rover in its place. After that, Amitch started running a bunch of calculations, and again, eventually dismissed that theory as well. This went on for about 20 minutes. Finally, Amitch paused for a about thirty seconds. Then, the Amitch system suggested that the most logical explanation was that those brilliant NASA engineers must have developed some sort of special permanent brake system. Further, they stated that this system was attached to one of the tires. To avoid being detected by the cameras during the different scenery angles, Amitch suggested the braking system would have to alternate between tires. If that wasn't the most farfetched idea you have ever heard, the Amitch system was a little buggy in its development stages.

Suddenly, Amitch took off again. They frantically searched the data bank of Apollo Moon Landing photos. They attempted to find some type of connection. Pictures were scrolling across the screen so fast that we could not see what they were. Then all of a sudden, Amitch stopped and the message "Answer Found" showed up on the computer screen. Boy was everyone excited, "answer found" was a special response code one of the Whizkids programmed into the Amitch System. This message was only supposed to show up when it established a final conclusion to a given problem. We had never seen that come up before.

A few seconds later, these next two pictures popped-up on the screen with the message "permanent tire brake system found". There it was. To confirm the moon landing hoax theory once again, this was the answer we were looking for. According to Amitch, the first picture was some of NASA's brilliant engineers back in the 1960's working on a portable braking system. The braking system was designed to prevent the rover from constantly moving around and contaminating the movie scenes moon dust.

10075959

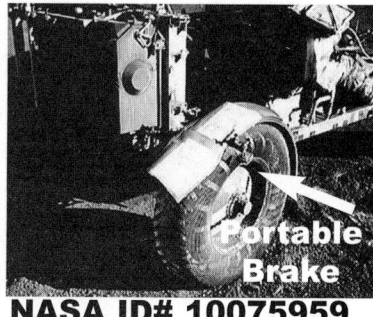
NASA ID# 10075959

Supposedly, they were discussing the idea of adding a shielding between the portable brake unit and the fender of the Rover Vehicle. The purpose of the shield was to prevent the brake unit from scratching the Rover Vehicle fender. Amitch came to this conclusion based on the fact that it could match scratches on the rover vehicle fenders to the portable braking system, which was attached to the Rover Vehicle. This can be seen in the picture on the right.

The close up picture of the moon rover vehicle with the attached portable brake was supposedly taken on the moon surface during the Apollo 17 Mission. Those NASA engineers were good. The portable brake was small enough to have gone undetected and certainly would have prevented the Rover Vehicle from rolling down the steepest hills.

Now that it has been found, how can NASA deny its existence, or suggest the astronauts could have ever driven the rover with the portable brake unit connected?

Could the Amitch program be wrong about this braking system? Of course not. Amitch theories have never been wrong yet, so why should this be different? Look at the next picture Amitch matched up with these other portable brake portraits. NASA claimed this picture was taken during the Apollo 17 Mission. The brake is attached to the tire, and there are no tire tracks leading to, or from the Rover Vehicle.

As seen in the picture to the right, this brake was detected in several pictures in the middle of a wide open area. After discovering this portable brake system, NASA now refers to it as a temporary fender fix. Since it is clearly a custom brake system and connected directly to the tire, that claim is ridiculous! Once the rover attempted to move, it would have ripped the fender right off. Besides, if this was some type of fender fix designed by NASA engineers, why would they design one that would prevent the Rover Vehicle from moving?

Braking System Attached to same to tire at different location again no tire track leading in either direction.

**NASA ID# as17-142-21730
Tire Locked With Brake**

**NASA ID# as17-146-22345
Tire Locked With Brake**

Methods Used To Fabricate the Moon Landing Scenery

The Pictures Showing No Footprints, or Tire Tracks Provides Another Clue to NASA's Moon Landing Hoax

Another strange thing about the alleged moon photos was the picture the astronauts were taking of the "Moon Tools". At first these would appear normal since NASA claimed the tripod was a market, which the astronauts were supposed to place next to rock samples they were going to collect or photograph.

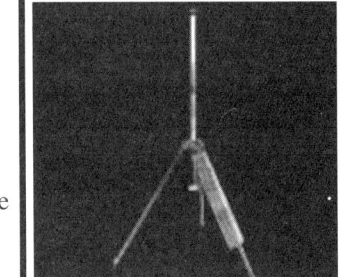

The three -legged device is a "Gnomon Color Chart"

However, there are many clues indicating the "Gnomon Tripod" was nothing more than a stage prop for NASA's Moon landing movie. Some of the clues are so obvious it's a wonder. After all these years, no one has noticed the missing footprints leading to and from the tripod.

NASA movie producers were using the "Gnomon Tripod" as a stage marker. That right, the movie creators primarily used the "Gnomon Tripod" to let the astronauts know where to find certain items hidden under the moon's surface, and also where to take certain pictures before proceeding to the next section of the movie platform. This is illustrated in the next set of pictures.

Look closely, and you will notice how the astronaut's footprints often ended just before the "Gnomon Tripod" or "Core Sampler". Hence, prior to the astronauts (actors') arrival, the movie producers placed the tools at the landing sites. Remember, this was also what they did for the Moon Rover Vehicle.

20134558   20134693   20134691   20148226   20148223

If these pictures were actually filmed on the moon, thjey would make no sense. However, when you think about it, if the pictures were taken at the Apollo astronaut training simulators here on earth, they would make perfectly good sense. It would have taken the stage crew hours to setup each new scene. Therefore, before moving on to the next filming location, it was critical that the astronauts knew exactly where to stop and take certain pictures, or they would have contaminated the next scene.

Now of course to say something is one thing, but to prove it is another, so additional evidence would be needed. If the moon hoax stage marker theory was correct, for it to be considered a scientific fact additional proof would be needed. Of course it should not be too difficult to find evidence that the moon landings were faked. It is very easy. Just go to NASA's website. The hardest part has been finding just one piece of evidence that proves the moon landings were real.

There are no footprints around the stage markers, and after the filming, there are footprints. This condition could not have existed on the moon. The pictures must have been taken from the Apollo training missions. It is the only thing that makes sense.

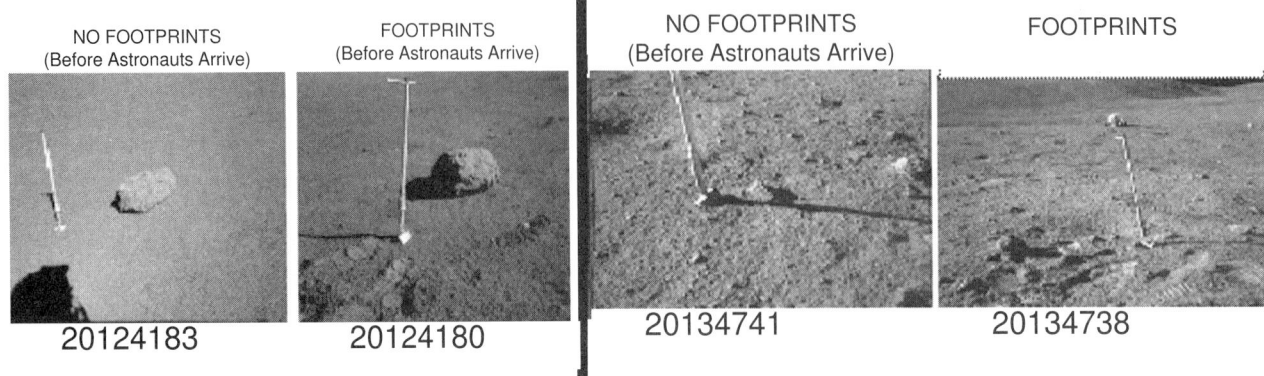

NO FOOTPRINTS (Before Astronauts Arrive)   FOOTPRINTS (Before Astronauts Arrive)   NO FOOTPRINTS (Before Astronauts Arrive)   FOOTPRINTS

20124183   20124180   20134741   20134738

These next picture also show of how the moon landings were an elaborate hoax staged here on earth.

The "Gnomon Tripod" provided additional proof that the American moon landings were a hoax

The "Gnomon Tripod" stage markers also provided evidence suggesting that, to produce the fake moon photography, NASA had been doing a great deal of cutting and pasting with different photos. In many pictures, the "Gnomon Tripod" matched the Apollo training missions to the identical locations on the moon.

That is right. The cut and pasted earth pictures appeared to be an identical match to the moon. It's just a matter of lining up the "Gnomon Tripods" on earth to the Gnomon Tripods on the moon, and the pictures match up. It's that simple. When NASA came up with the idea for using the "Gnomon Tripods" to help match photos being merged together, it was a brilliant plan. Back in the 1960's, how would NASA know someday humans would develop artificial intelligence with the capability of matching these photos and so much more? In the past, the smartest computer on the planet had less computing power than a small calculator or wristwatch of today.

NASA ID# ap17-KSC-72PC-379 -KSC

NASA ID# 10075952

When matching up the earth and moon Gomon Tripod on stage markers, here are a few examples of the types of results obtained. The results are exactly what one would expect. These matching set of pictures illustrates how the moon dust sprayed over training simulators, and cut and paste photography editing techniques used, which will both be discussed in detail.

To create the above picture on the right from the picture on the left, NASA movie crew cover the rocks with some artificial moon dust. Then took a picture from behind the rocks by the trees, and you have the moon picture seen on the right.

10075927

Samples of NASA Photographers using a Cut and Paste Technique

Amitch matched these groups of pictures as being used to cut and paste into one moon landing photo. All NASA would have to do is fill the earth picture on the left with moon dust. After that, paste the astronaut picture inside the Kennedy Space Center over it, and you have an identical match. Why would NASA ever go through the trouble of sending men to the moon when they clearly had everything they needed to fake the moon landings here on earth?

Advanced "Cut and Paste" Technique

The two pictures on the left were used to create the moon landing picture on the right. Under close examination, after removing the airbrushed changes, the picture on the right is the same as the moon surface in the middle picture.

ap15-s71-37218HR   NASA ID# a149243   NASA ID# a149241

Mysterious pre-made holes found in the moon surface confirms Moon landings were fake

In order to convert any good theory to a scientific fact, it needs to be open to independent examinations. With all the hard work involved in making such an important discovery, it would be a shame not to make this evidence available for independent operations and discussions. Besides, in many cases, this type of interaction can often reinforce a solid theory such as the American moon landing hoax contained in this book. Whether right or wrong, no one should ever be held back from expressing their opinions. This whole book is just a friendly discussion and exploration of possible theories regarding the controversial moon landing evidence.

Unfortunately when it comes to evidence and theories in an open discussion, NASA officials don't see eye to eye. Why else would most of NASA's evidence related to the Apollo moon landings be sealed in a room until the year 2026? Why else would NASA have 24 hour armed guards blocking the door to the storage room with orders to shoot and kill any intruders? Something is not right! Now, what kind of scientific evidence do you think might be hidden behind that door? Well, for those of you that live to 2026, surely based on the limited evidence already available to the public, you're going to be in for the surprise of your life.

For those of you who can not wait until 2026, your about to get a big sneak preview of what is to come in the future. The following proof of the Moon landing hoax is the most compelling evidence of all, and it was a direct result of a simple question. The students, teachers, and scientists from around the world who worked on this moon landing hoax project were all asked these questions. What else needs to be answered to convince you that the moon landings were for sure fake? Can you think of anything else that could contradict the moon landing hoax theory contained in this book? Is there something not yet answered, or that does not make sense about the evidence presented here as proof the moon landings were fake?

Apparently the evidence in this book is more than originally thought. There was only one last doubting question remaining. If this remaining doubt could be answered, then the evidence contained in this book should be able to confirm the Moon landing hoax, rewrite Earth's history books, and correct this error in time.

10075189

These people are absolutely right. The custom American Flag theory had a hole in it and everyone spotted it. The stage prop flags would have tipped over every time it was placed in the soil. Unless some type of device was either supporting the stage prop flag from above or below, how could this be? Nothing could have been supporting the flag from above, or it would have already been spotted by someone. What about under the moon surface? To support the custom stage prop flags, could NASA movie producers be hiding something under their artificial moon surface? If history repeats itself, there should be a very logical explanation supporting the Moon hoax theory.

It is a little strange that the American Flag was usually found standing perfectly straight up. This indicates that the lop-sided American flags were inserted into some type of holding bracket on the surface. Since the core sample driller would have also needed a long tube, we are looking for a system that could also extend several feet under the ground, most likely a long tube. All we have to do is try to find some type of small-premade holes that were camouflaged on the Moon surface, which should not be that difficult. From what we can see in the Moon landing film footage the astronauts were able to find these pre-made flag pole holes and they could barely see the ground themselves. Since the place was very dark, they were nearly blinded by the powerful movie studio lights beaming at them.

Realizing NASA would have to use some type of system to direct the blinded astronauts to the pre-made holes. All we had to do was figure out what was being used as the stage marker to direct the astronauts to the location of the camouflaged flag pole support brackets.

PRE-MADE Flag Pole Holes Found

You guessed it; the pre-made holes were found for the flagpole and core sample drilling films. Look closely at the next set of pictures below, and you'll see that old "Gnomon Tripod" served another purpose. Notice how the "Gnomon Tripod" was pointing in the direction of the pre-maid holes in the stage platform. Some photos without any footprints leading to, or from, the pre-maid holes. How convenient. Apollo Mission Portraits show the moon was already equipped with custom holes for the flagpole and core sample driller. Those Actornauts must have loved that!

With a magnifying glass, you can easily see the pre-maid holes were camouflaged with what looks like plaster and artificial moon dust. These pre-made holes were hidden in the exact same way the plumbing pipe and electrical connectors were under the LM spacecraft. Could anything be easier than that? The "Gnomon Tripod" tells us exactly where to look for things that NASA movie producers were hiding from us in the photos.

Pre-made Flag Pole Holes Were Camouflaged

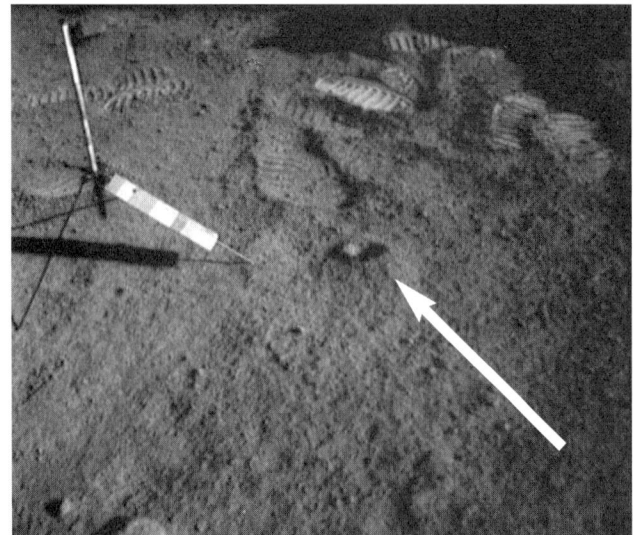

NASA ID# 20147736
CS Spring Hole

No astronaut footprints lead to the Pre-made flag pole hole
NASA ID# 20147706

More Pre-maid Holes Found Camouflaged On Moon Surface

NASA ID# 20147721

NASA ID# 20147733

The Flag Pole Bracket Assembly (CS Spring).

The Amitch System even provided the picture of the actual flagpole bracket NASA referred to as the CS Spring. As seen in the picture below on the left. The plastic rings were used to insure the flagpole would slide in a relatively easy fashion.

To insure it would not be detected, the middle picture shows how the bracket was camouflaged. Now the picture on the right appears to be one of NASA's engineer's failed attempts at figuring out how to get the American stage prop flag to stay up on the artificial moon surface. Propping it up with moon rocks was obviously not one of there brightest ideas. Before they invented the CS Spring System, one can just imagine how frustrated they must have been.

judy17.jpg

NASA ID# 20147721

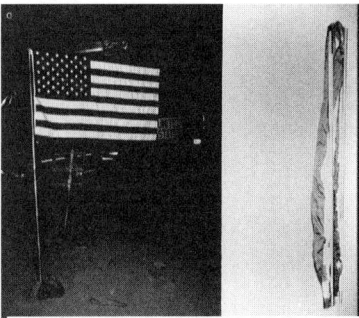
NASA ID# 10075189

If you're saying to yourself, if this Amitch System was so smart, why didn't it come up with the actual words for the letters CS? It did. They are called Compression Sleeve Springs. If you don't believe it, ask the NASA officials. They should be able to verify it. The Amitch System was so smart it even determined every location of a spring. A CS Spring had its own special number. For example, the pre-made hole for the American Flag to be inserted during the Apollo 11 mission was referred to as CS Spring #2.

For the skeptics that have a hard time believing this theory, wait until you see this next piece of evidence the Artificial Machine Intelligence Technology Computerized Hyperthinking system provided during its last transmission before the program finally crashed for good.

The Astronaut's live action footage of this Apollo 11 flag mounting scene revealed exactly what would be expected using this type of flag pole holder. In this video, the astronauts were seen having a very difficult time finding this pre-made hole and inserting the flagpole into it. In the alleged moon film, it was clear that the reason the astronauts were having so much trouble finding the CS spring holes was because they were being blinded by the stage lights. As stated before, the lights were pointed directly at them. The astronauts could only get in close proximity of the hole. Because of this, they had to search around until they found it. This live action flag pole filming is located under the NASA Apollo film number a11f.110085.

NASA ID# a11CSspring2

NASA ID# a11f1101057

While watching the Apollo Moon landing films, one can see the astronauts are purposely kicking moon dust over the base of the flagpole and driller. This was an attempt to help keep the brackets that were earlier embedded into the artificial from being exposed during filming. This is especially noticeable during the alleged live broadcast. If the brackets did not remain covered at all time it would have revealed to everyone NASA was faking the moon landings.

Methods Used To Fabricate the Moon Landing Scenery

When NASA is confronted with this group of composite photos proving the Moon Landing scenery is fake, they will very likely feel foolish and quickly admit to faking the whole thing.

Below are the filled in crater photos mentioned briefly earlier, that reveal NASA's photos of different Moon landing site the same background mountain scenery. Even though they were supposedly on different areas of the moon. It appears NASA used the same scenery in many different landing site photos. If it were anyone besides the American government behind these photos, there would be no question this photography is 100% fake.

Filled in Crater Confirms NASA was Using Composite Pictures

These next two photos with identical backgrounds, hold many hidden secrets to NASA's photography deception. First, there are supposedly two different locations on the moon surface. These locations are Station 4 and Station 6. Station 6 shows a mountain scenery with a large crater in front of it. Now look at the Station 4 photo, it has the same mountain scenery, but the crater has now been filled in. How can this be?

Two totally different Apollo 15 Moon landing sites with what looks like the same background scenery, how could this be? Under close examination, it's clear they are both pictures of the identical background scenery, taken from the Pancake Range area of south-central Nevada.

In the first picture, there is a large crater in front of the hill. This is believed to have been used in many NASA moon pictures where the background scenery was alternated between scenes.

Additional proof these photos are fake can be seen in the second picture. The crater is now filled in with small pieces from other pictures. This is confirmed by the large number of mismatched crosshairs in the filled in crater area.

16-24

More Compelling Evidence Of NASA Using Trick Photography

Up to this point, there have been many examples showing how NASA's photographers used a variety of manipulation techniques to produce the fake moon landing photography. By now it's obvious that the photographs were faked, and at this point, any additional evidence is likely to be considered an overkill.

However, it's important that all evidence related to the moon hoax was made available to the American space agency. This will insure that, when they prepare their explanation of how the moon landings were fake, no topic will be left out. So this chapter will provide additional evidence to the photography discrepancies, and prove the moon landings were phony.

Some of the evidence can be considered particularly disturbing, however each image is in its original form. When possible includes an official NASA identification number for verification, other than sections enlarged for illustrative purposes.

Proof Of How NASA Used Their Artists To Help Construct the Fake Moon Pictures.

This next set of pictures show how the NASA photographers and artists worked as a team to make the realistic looking Moon landing photography. These three images were supposedly taken during the Apollo 16 mission, however there are many clues revealing they were taken on earth inside the Kennedy Space Center.

The image on the left shows the beginning stages of a moon scene construction. At this point, the photographers added a flag and an astronaut to a picture previously taken at the Kennedy Space Center Training Simulator. The picture in the middle reveals how, by adding footprints under the astronauts, the artists were beginning to touch up the surface. The picture on the right shows the footprints and final rock formations were completely filled in by the artists. Notice how the astronaut's feet are partially cut-off as a result of the Artist doing a poor job of matching the astronauts to the moon surface picture. Also the artist has clearly mixed up the original rock formations and the final picture on the right is a lot different than the first two originals.

No Footprints
as16-113-18339

Footprints Added
as16-113-18340

Different Rock Arrangement
NASA ID# as16-113-18342

More Samples Of Composite Pictures:

The next set of visual renderings were obtained from NASA. Furthermore, they reveal in more detail, the cut and paste procedure used to create the realistic looking moon photography. There are many obvious clues this Apollo 15 scene was pasted together. The first clue is the scenery was obviously constructed using pieces of many pictures to distort the moon landscape. Next, the 17,000 ton LM spacecraft changed its location 3 times in the different NASA photos. In the first composite picture, the spacecraft was leaning to the left. However, in the second photo, the spacecraft is leaning to the right.

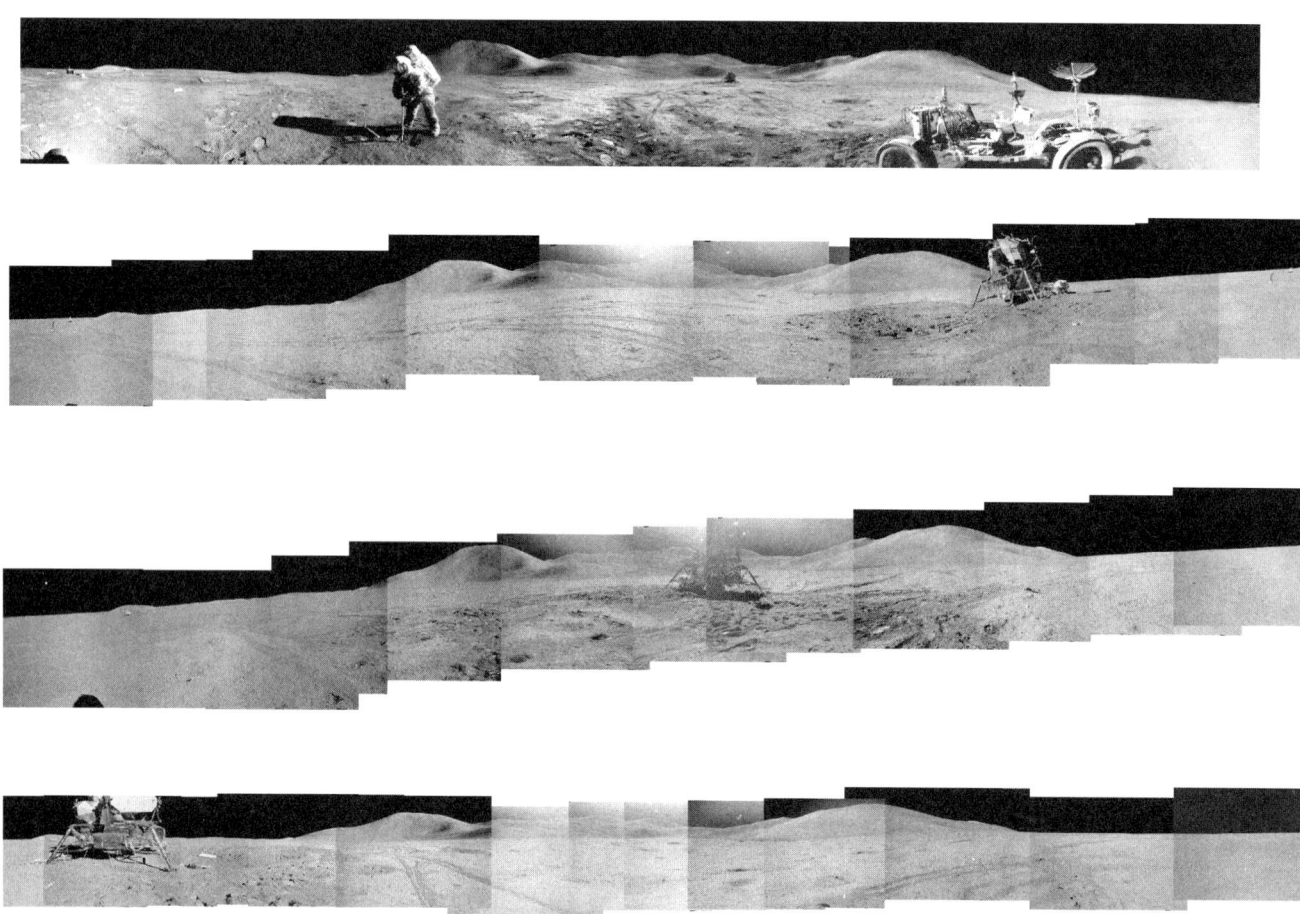

NASA could only attribute these photos as an attempt to create moon landscape scenery. Illustrations with anomalies like this could not have occurred on the moon. NASA's Apollo 15 mission records indicated that the spacecraft never changed its landing position.

Amazingly, this is not the only adjustment NASA photographers made to the Apollo 15 landing site. As shown in the series of NASA composite pictures below, they also added craters, gullies, and hills.

Foreground Changing.

The Entire LM Spacecraft Disappears.

The next set of NASA composite pictures not only show the photographers have changed the foreground landscaping, but also the LM spacecraft has disappeared.

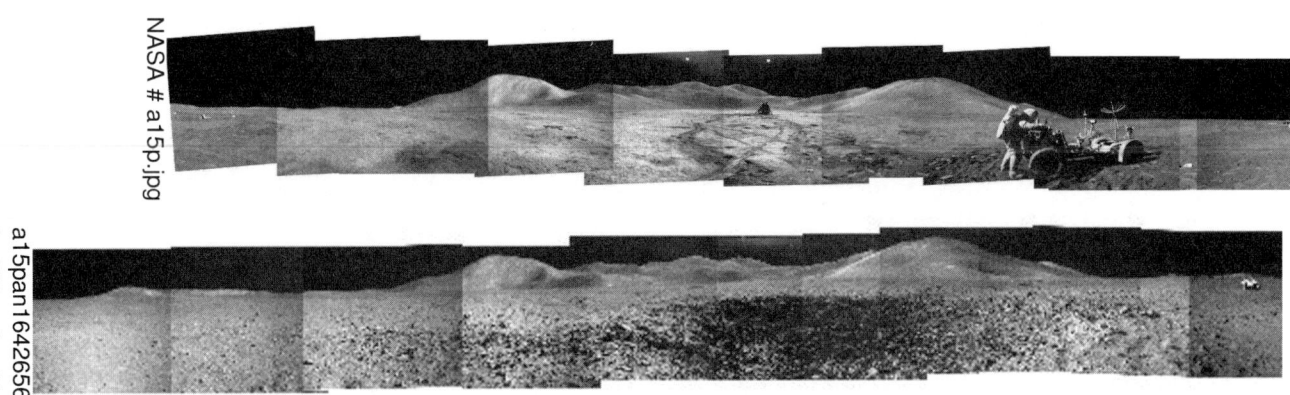

CHAPTER 17

NASA Artists and Photographers Leave Clues to Moon Landing Hoax

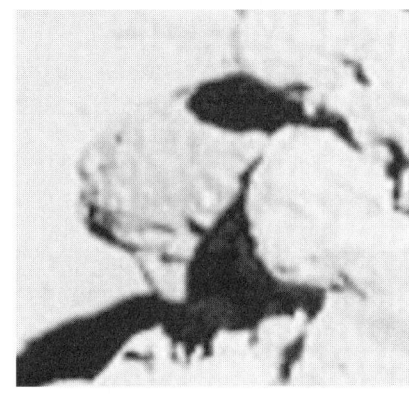

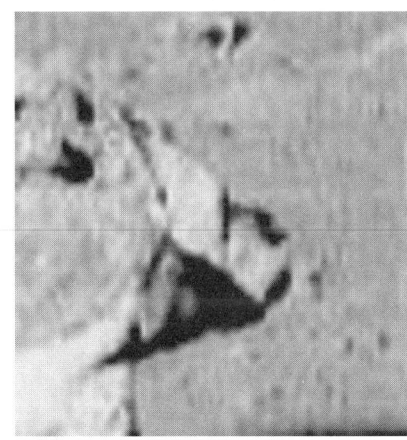

Chapter 17
NASA Employees Leave Clues to the Moon Landing Hoax

Like most good conspiracies, there is always the possibility of someone working on the project, leaking information, and NASA's moon landing hoax was no exception. The evidence presented in this chapter shows how NASA had a big problem with employees attempting to leave clues. Furthermore, before releasing any evidence to the public, they had to create a Pre-screeners Department to examine it. The Pre-screeners were NASA's watchdogs, and they viewed every photo. The watchdogs looked for any clues left by employees.

This was an enormous task for NASA's pre-screeners since there were an infinite number of ways someone could leave a clue, which could have revealed the secret. If the slightest thing had gone undetected and was released to the public back in the 1970's, it would have easily exposed NASA's moon landing hoax. In the past, many people were skeptical, and were looking for the smallest clues of NASA's wrongdoing. However, the American Government has been brainwashing people into believing the moon landings were real for so many years that time has reversed their thinking. Presently, most people assume the American moon landings were real. Because of this, folks ignore any opinion to the contrary.

NASA ID# 20130756

Since so many years have passed since the alleged Apollo Moon Landings, it's difficult for the human mind to think something is wrong with NASA's proof. Today, if NASA releases a photo containing clues that the moon landings were faked, it simply goes undetected. For example, people are so convinced the moon landing pictures are real, they instinctively overlook the fine details. For example, in the NASA moon landing picture above, there is a big dog sleeping right in front of everyone's eyes. This picture has been around for some thirty plus years and nobody has ever seen the dog?

Yes, "A dog" crawled onto the Langley Research Center's outdoor moon simulator and took a nap. If you do not see the dog yet, don't feel bad. With NASA's custom light filters attached to the movie cameras, it was highly unlikely anyone would recognize the dog at first. Unless someone had told them what to look for and pointed out exactly where the dog was sleeping, nobody would have noticed the animal. That's right, the dog was found after discovering the method NASA's pre-screeners used to mark problems with all the moon landing photography. Everything from animals getting into the pictures taken at the outdoor studios, stage props being exposed in pictures, even NASA employees leaving clues in an attempt to tell everyone the moon landings were faked.

Understanding NASA's photography pre-screening system helped explain why there seemed to be so many random clues that made no sense. There would be a discovery of one type of clue, but not quite enough to prove anything. Then another totally different set of clues would be found. Again, the hints were not enough to be considered solid proof of the hoax. However, when the clues were analyzed as one group with the pre-screening, there individual meanings were revealed.

It appeared many NASA employees left clues. None of the workers were aware that the other employees were also trying to leave clues. As you can imagine, each NASA employee had to be very secretive about what they were doing, and for a good reason. The American's moon landing hoax was some very serious business. If you were a NASA employee caught attempting to leave clues about the moon landing being faked, most likely the CIA would have killed you since you could never be trusted again.

Because each person was working on their own, it explained why there were so many different types of clues, but no significant pattern of solid proof. Each person working on the moon landing project had different access to project materials, which determined the method they could leave their clues. For instance, several artists, working on the project, resorted to altering moon landing pictures slightly. The stage crew and astronauts left secret messages and different objects as clues because they wished for them to be found.

What is kind of interesting is the NASA employees, who left the clues, were all trying to get the same message across in there own special way. The message was "The moon landings were faked". It's not surprising to see they all attempted to use the same method of message delivery. The media was certainly not an option. Can you imagine the Apollo Astronauts, Pete Conrad or Buzz Aldrin, trying to sell this story to the National Enquire?

Back to the dog picture, that was discovered from a clue left by NASA's pre-screeners. The portrait below is an original unaltered NASA photo with ID # 20130804. This picture marked the dog's location. This made it possible to see the same dog in other NASA pictures from a different angle. In fact, this dog was seen in many of NASA's alleged moon landing pictures. It's believed that the animal was a pet of one of the astronauts or movie producers. This has not been verified yet. If you look closely, someone working for NASA left a clue. This object was a dog. Like many of the clues, it was very simple and right to the point. The letter 'A', and a picture of a dog.

NASA Leaves Clue That There Was "A dog" On the Moon!

Although the dog in the picture on the right is impossible to fully make out, there are other much clearer examples that have been found. In the moon landing photos, this dog can be seen in many different locations. It's fur is blended in so well with the color tone of the artificial moon surface, NASA pre-screeners often overlooked the dog completely.

Below are a few more examples of the moon landing images with this dog.

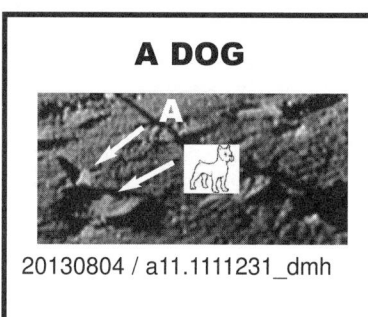

Dog's First Sleeping Location on the Moon

NASA ID# 20130756- Dog Sleeping

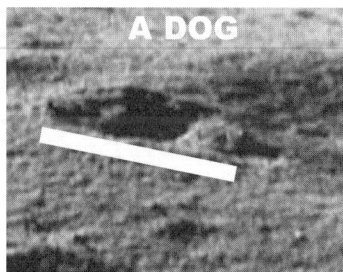
A Dog Sleeping

Dog Moves To A Different Area

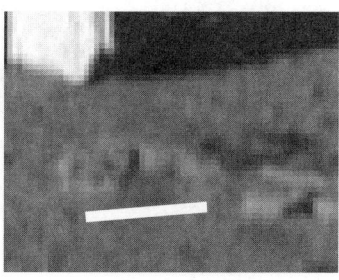

A perfect example of how NASA's pre-screeners used these white objects to mark problems with certain photos would be the picture of the Astronaut Buzz Aldrin photographed inside the Gemini 12 spacecraft cabin during the space flight. This picture was taken on 11/13/66.

In the center of the photo, You'll notice that there was a white marker placed over the actual original photo. This white marker was placed there by one of NASA's photography pre-screeners to indicate that, while supposedly navigating back through earth's orbit, the astronaut had been smoking his tobacco pipe. If this picture was really taken in outer space, that would have been suicide.

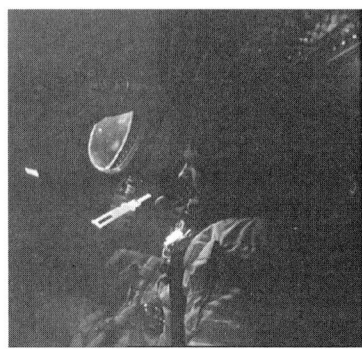

Astronaut Smoking Pipe
NASA ID# 10074594

Discovering the connection between the pre-screeners white markings and how they pinpointed the exact location of the actual problems found in each picture, unleashed an abundance of new evidence related to NASA's cover-up. Throughout this chapter, you'll notice all kinds of new and exciting clues to the moon landing hoax. These white markers revealed the location of many different types of objects on the moon including more dogs, cats, people, toys, and much more.

This next picture is of a cat. Thanks to NASA's pre-screeners, this cat was one of the first of many animals to be discovered in the Apollo Moon Landing pictures.

A Cat Found On the Moon

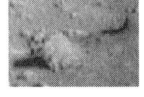

Siamese Cat On Moon
NASA ID# 20124287

A Barbie Doll Left By A Little Girl Playing On the Movie Studio

Thanks to NASA's pre-screeners leaving white markers, this Doll was also discovered on the moon surface. It looks like a person. Like the dog, the reason the white marker was made to look like a person was to let the touch-up artists know what to look for. Additionally, it educated the artists on where the objects were positioned.

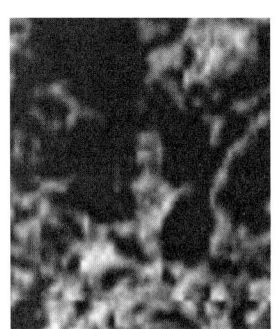

Child's Toy Doll On Moon
NASA ID# 1007528

By far, the NASA artists were the most creative people in coming up with ways of leaving clues about the moon landing hoax. It appears a few of the artists, hired by NASA to touch-up the moon hoax photos, decided to risk their lives by adding clues into the moon landing photos. To the naked eye, many of these artist's clues were not so obvious. However, when considering the uniqueness of the artist's clues, they can only be considered deliberate attempts to leave hints.

This next illustration shows how creative the artists were in leaving clues. Theses artists would form all kinds of objects out of moon rocks. This is a half man/half rock figure. It appears the man chiseled the word "FAKED" into the side of the rock in front of him. To the left of the half rock man is a white stick-man marker left by the NASA pre-screeners. This marker is pinpointed to the rock man's exact location.

NASA Artist Creates Man Out Of Moon Rock

NASA ID # 10075064

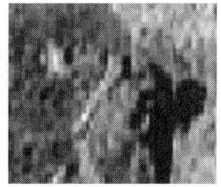

Maybe this is Moses finishing the 11th Commandment
"Thou Shall Not Lie"

The presence of the white stick man marker indicates NASA's watchdogs also discovered this clue. Hence, we need to add suspicious NASA Artist's deaths to the list of suspected CIA murders related to the moon landings. There were so many examples of this NASA artist altering the appearance of moon landing rocks. Why did she do this? It makes no sense. Sooner or later, NASA's watchdogs were going to catch her. If she had survived, there was no telling what she could have done. Most of her masterpieces would have made Da Vinci and Buonarroti proud.

The good news is NASA accidentally released some of the artist's best work. It appears the pre-screeners were unable to catch every one of her signs because they blended so well into the scenery. This next artist's clue is a perfect example. It stared everyone right in the face for over thirty-years, and no one has seen it.

Where Is Waldo? He is On the Moon!

The head, on this rock, is believed to be this artist's boyfriend Waldo, and proved the clue to the famous riddle, "Where is Waldo". The mystery is solved. He is on the moon. Look, he is hiding behind that moon rock.

Human Head
NASA ID# 2012989

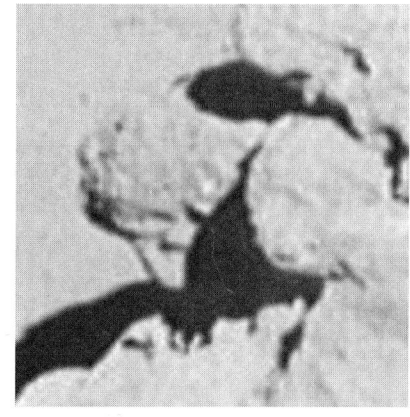
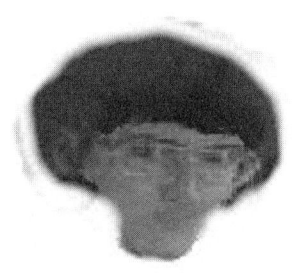

Rat Rock:

It is hard to tell if this is one of the Rock artist's clues, or a real animal the astronauts came across in the desert. It does not really matter if it's on Animal or a painting, either way it was obvious that the photo was not taken on the moon.

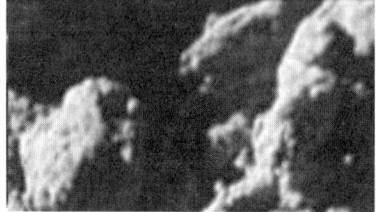

Animal in rock
NASA ID# 20121981

Man's Best Friend Rock

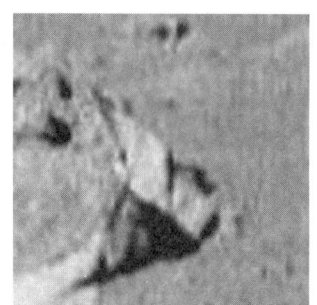

Dog Faced Rock
NASA ID# 10075629

Apollo 12 Astronauts Narrowly Escape Moon Monster Attack

There were many rock formation clues left by NASA artists and sometimes they got a little carried away. Please remember this is a NASA original photo, which they claim was taken on the Moon. After seeing this picture ask yourself if anything NASA claims about going to the Moon could be legitimate. The picture shows a snake holding the Apollo spacecraft in its mouth while supposedly on the Moon during the Apollo 12 Mission.

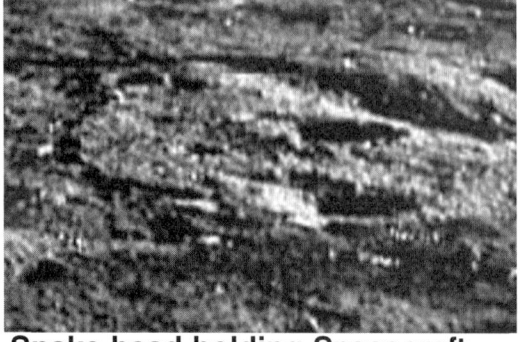

Snake head holding Spacecraft
NASA ID#as12-46-6737

If you are beginning to wonder why NASA had released these pictures to the public, that's a very good question. It was simply impossible for NASA's watchdogs to recognize all of the clues that were left behind by employees. There were several different artists leaving such a wide variety of clues, it was impossible for the pre-screeners to know what to look for. Also, some of the evidence was so strange, it was difficult for the human brain to understand them at first. This made the watchdog's job even more difficult.

"You Have Been Buffaloed"

If this artist's next clue had been discovered at the time of the alleged moon landings, most people would have understood it. This picture illustrates the Great American Buffalo. In the English Language, the word "Buffaloed" referred to something being faked and was used commonly in the 1960's. The word Buffalo was used by itself. It means something was fake. It could be used in the phrase "You have been Buffaloed" meaning someone has been tricked. This was what the artist was trying to let everyone know with this moon landing picture alteration. It's the backside of a Buffalo with its head facing left.

NASA ID# a15.1455513_dmh -From Back Side, A Buffalo Looks To the Left

The Flying Goose Head.

Again this goose head was deliberately painted into the picture by one of the artists working on NASA's moon hoax as clues for people to find.

NASA ID# a11_5847-8_dhm

The Monster Rock:

Not sure what to make of this funny looking moon rock, but If this rock could talk it would probably tell a lot about how the Moon hoax movies were really produced. We'll call this Moon rock photo the "Producer's Favorite" because it looks a lot like Jabba-the-Hut from Star Wars. Under close examination we can see the astronauts and stage crews reflections on the film as if they are looking at the large monster rock and singing a song about the pair of two funny eyes they are looking at. Those are the words the astronauts are really singing as they head for this rock, with a large tongue and a pair of funny eyes. The astronauts were leaving everyone a clue to the moon landings being faked.

NASA ID # apo16

Below is most likely the mirror reflection seen in Monster Rock Photo above

If you still don't believe the moon landings were faked, please don't look at the pictures on this page. It could be very scary. The NASA artist who left these clues was into space aliens and left the strangest clues ever imaginable. NASA never discovered them because they blend in so well. These moon landing images contained space alien babies, dead aliens, alien pets, and alien graveyards. If these pictures were indeed taken on the moon, everyone should be scared for their lives. It's just a matter of time before space aliens are going to be coming to take over planet earth.

NASA Artist Leaves Evidence Of An Alien Race Living On the Moon

NASA ID# 20130746

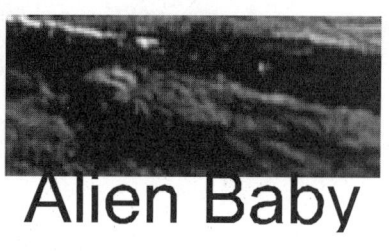

Alien Baby

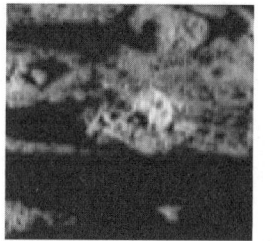

Alien Skull Found On Moon

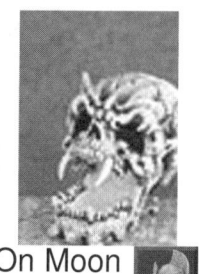

This next picture shows one of the Apollo Astronauts discovering an alien gravesite. It appears he is marking the location for future explorations. Decomposed alien bodies can be seen all over the place.

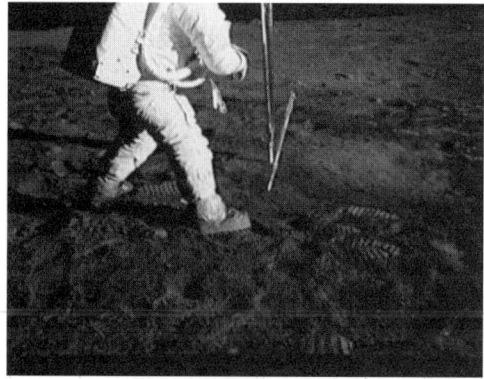

NASA ID# 20130806

Astronauts Discovered Alien Graveyard On the Moon

Alien Pet

NASA ID# 20130794

Alien's Pet Dog Run Over By Astronauts

Alien Horse Lying Next To Rock

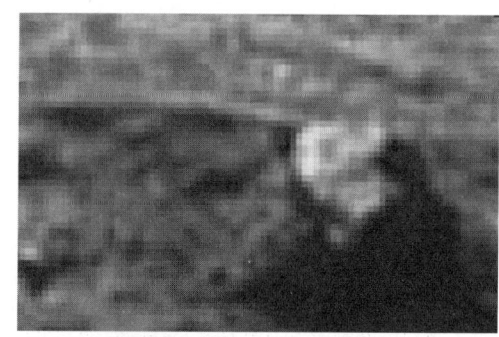

NASA ID# 20130755

NASA Artist Paint Clues In Helmet Reflections

Another NASA artist painted clues in helmet reflections. Below are a few visual renderings.

White Rabbit Reflection On Helmet

NASA ID# as17-136-20760
White Rabbit On Helmet Reflection

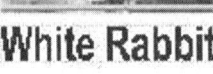

Fish Reflection On Helmet

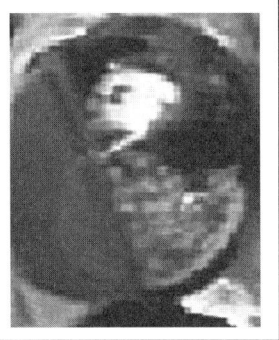

Is A Dinosaur Staring At the Moon?

This is one of the more obvious helmet reflections showing that NASA artists often painted the reflections on to the astronaut's helmet visors.

Some of the younger WhizKids, helping with the research, felt the reflection in this helmet resembled a dinosaur looking at the moon. Evidence suggests that because of the artist's poor effort, it's difficult to make out the detailed features of the equipment. In addition, as previously stated, because of poor attention to detail, the reflection in the helmet was backwards. If the astronaut was facing the sun, as this picture suggests, the object in front of him would be completely black not brightly illuminated as seen here.

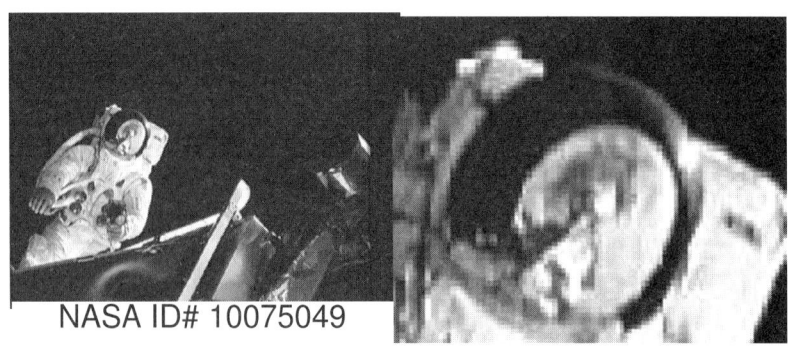
NASA ID# 10075049

The Hiker And His Dog;

Of course it is not possible for a man and his dog to be walking around on the moon. Furthermore, there is a good explanation for this unexplainable phenomenon.

NASA ID# 10100194
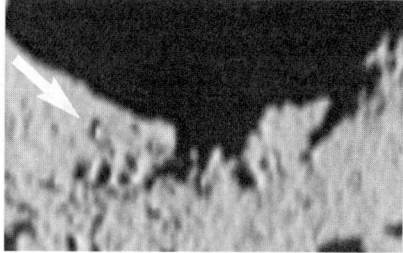

As mentioned earlier in the book, NASA often produced moon landing sceneries by cutting and pasting many pieces of different desert scenery taken from planet earth. For the most part, these photos were of various Apollo Astronaut training locations.

To avoid the possibility of anyone recognizing the earth scenery being used in the moon hoax photos, before piecing them together to make totally new scenery, NASA photographers dramatically reduced the magnification of the various scenery. When thousands of pieces of desert scenery are collected from around the world, shrunk, and mixed together, no one is able to recognize any of the locations.

NASA was correct. No human would be able to recognize the surroundings. However, they neglected to realize that reducing the scenery to such a small scale would distort certain items. To illustrate, see the above NASA moon landing photo.

After examining the alleged moon landing photos closely for more evidence supporting this cut and paste theory, many more similar items, such as miniature humans were found. Below is another example of a NASA moon landing photo showing a miniature astronaut on the moon. What was NASA thinking? If you had paid closer attention, you might have never been caught.

Miniature Astronaut Only About 3 Inches Tall

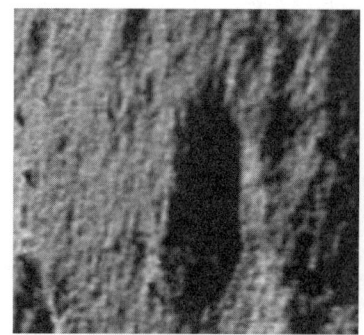
Miniature Astronaut
ID# 20130756
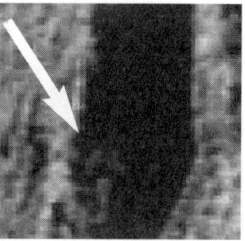

President Nixon Wearing Monkey Mask From Planet Of the Ape's Movie Set

Is there anyone in Hollywood that didn't know the moon landings were fake? This is more evidence proving the American government controlled their news media.

NASA ID# 10075309

10075311

One of NASA's employees intentionally pasted a monkey's head over President Nixon's head, and somehow for many years, it has been overlooked by NASA's watchdogs. Because at the time this book was written, this picture could still be seen on NASA's web page, one would have to assume NASA released this picture several decades after the Apollo moon landing missions. Certainly, if this picture were made available to the public back in the 1970's, everyone would have immediately understood the clue to the moon landing hoax. Nevertheless, over three decades, it's very unlikely anyone would understand its meaning.

When you release what was considered common knowledge to the people living during the Apollo missions, the message is quite clear. The monkey's mask was from a very popular movie airing at the time called "The Planet of the Apes". The story was about NASA astronauts that traveled to another planet. This planet was run by human-like apes. The monkey head belongs to Doc Sayous, who was the leader of the ape world. This relates to President Nixon because, at this time, he was the leader of our world. The show leads everyone into believing that the astronauts traveled to some other distant planet. However, at the end of the movie, the producers revealed that the astronauts were not on some other planet. The entire movie was staged here on earth and everyone was fooled into believing the astronauts were on another planet. The clue this NASA employee was trying to leave is simple: "The moon landings were filmed here on earth".

Learning How To Find Artist Clues in NASA Moon Landing Images

Because many Artists hired by NASA were secretly leaving clues at the same time no one realized how many clues they were producing all together. Combined they produced hundreds of Moon Landing photos and paintings containing clues regarding NASA's hoax. Most of these artist clues have not yet been discovered.

Below is an original NASA Moon Landing photo where one artist got a little carried away. This photo is filled with different clues. The Artist must have wanted to make sure these images did not go undetected as being a fake. If you want to test your own detective skills and also have a little fun, see if you can make out some of these artist clues regarding NASA's moon landing hoax. You can post your results at www.whizkids.tv or www.moonbloopers.com if you like and then you'll be entered into a drawing for a prize.

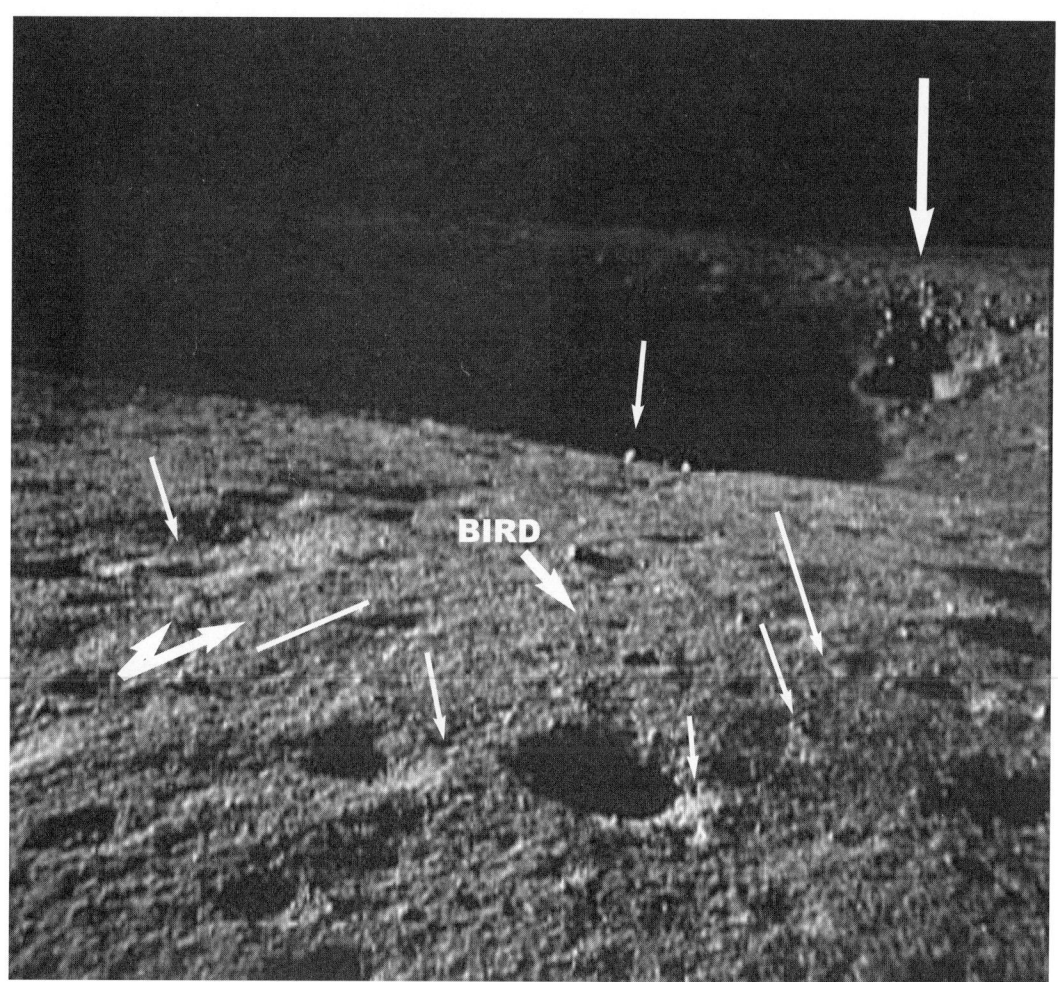

This Moon Landing picture is full of artist clues.
NASA ID# as12-46-6746

What Do You See?:

_____

_____

_____

_____

CHAPTER 18

The Moon's Animal Kingdom

Chapter 18
The Moon's Animal Kingdom

As we go through this chapter, you'll see that many moon landing pictures contain earth creatures of different types. Proving, the moon landings are nothing more than a complete hoax. We know the Apollo Spacecrafts were not large enough to transport all these animals to the moon. NASA would have needed something like Noah's Arc. Besides, even if NASA did manage to some how get these animals to the moon, there is no way any of them could have survived on the moon surface without some type of life support system.

There is a logical explanation for earth creatures found in the moon landing photos. It's exactly what would be expected. One would have to imagine that dealing with wildlife was one of NASA's biggest challenges. Good grief, because there are so many animals seen in the moon landing photos, you would think the filming was done at a petting zoo instead of on a fields outside the government space centers.

Displayed below is the first earth creature discovered in the moon landing images. It's believed to be a full-grown Siamese cat. The cat is looking right at the person taking the picture.

Siamese Cat on Moon Surface   NASA ID # 20124287

Why, in the world, did NASA ever release such a ridiculous photos in the first place? Could it be because the rest of the pictures were much worse than these? Was this the best that NASA could come up with? This would explain why most of the other moon landing photography is being concealed from the public. Additionally, the location had armed guards blocking the door 24 hours a day. Security officers are orders to kill any trespassers. What was NASA thinking? They left such incriminating evidence only protected by armed guards. That's extremely vulnerable to a security breach. How come NASA never moved the evidence to area 51 where they were hiding all the good stuff?

At first, a lot of the animal evidence was very startling, so we'll start off easy. If you would like to see these pictures in high resolution, they may be accessed at NASA's home page, or you can find them on the Whiz kid's website at www.whizkids.tvor www.moonbloopers.com

They have a lot of additional moon hoax items with much better resolution than the pictures contained in this book. Please feel free to check it out; you'll have the time of your life.

Chapter 18- The Moon's Animal Kingdom

Cat Trying To Hide From Astronauts In Moon Crater.

While filming the Apollo 15 mission scenes, a cat found its way onto the studio. Based on the cat's footprint distance, it appeared to be running. Since the cat was spotted trying to hide in one of the moon craters next to the rover, someone is most likely trying to capture it.

In the middle illustration, the cat is in front of the moon crater. He is looking back this way, as if someone were chasing it. Then in the next picture, the cat crawls into the moon crater and assumes a crouched up position.

Cat Discovered 10100258   10100259   10100260

Proof Of Animal Footprints

A white cat has found its way on to the moon surface and is leaving a trail of pawprints (footprints) in the moon dust. This 'white cat" can be seen in several photos by the Apollo 14 astronauts while they were supposedly on the moon. It appears as if the cat was first trying to catch the golf ball were hitting around. Then the cat ran around trying to get away from the stage crew as they tried to catch it. Eventually the white cat rested on one of the moon rock stage props as seen in the images below.

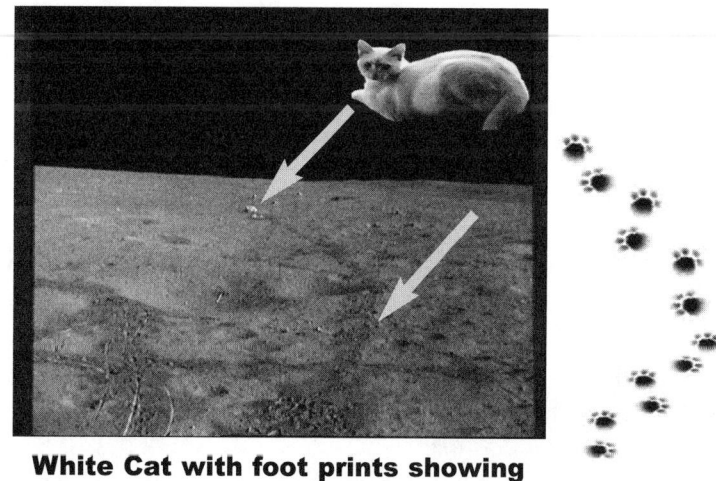

White Cat with foot prints showing
NASA ID# 10075613

White Cat ID# 20133922

White Cat ID# 20133921

18-3

Since NASA was filming outdoors in the same locations from 1963-1970 using an artificial moon surface, it made a perfect breeding ground for animals. Cats were especially attracted to the moon landing outdoor studios, it was the world's biggest litter box.

Prior to filming, to cover the animal's footprints, NASA quickly sprayed a thin layer of artificial moon dust on to the movie stage. This is evident, as seen in several moon landing pictures including the one here on the right just as one would expect, the thin layer of moon dust did not cover the cat's poop.

The moon landing litter box theory explained why an overwhelming number of moon landing photos have cats in the pictures. Below are a few more examples of such photos. If you wish to see more clear images, again go to www.nasa.com, or go to the Whiz kids web site at www.whizkid.tv

Cat poop found on moon.

More Cats Found On the Moon:

Evidence proving NASA knew about the cats being in the Moon Landing Photography

The Moon landing pictures between 10075600 and 10075617 contain Animals sleeping on moon landing equipment. These three images in the group have cats sleeping on the equipment. What was NASA thinking?

Cat Sleeping on Equipment
NASA ID# 10075617

Cat Sleeping on Equipment
NASA ID# 10075612

Cat Sleeping on right side
NASA ID# 10075611

Chapter 18- The Moon's Animal Kingdom

Cats Were A Major Problem For the Apollo Moon Landings:

Over the years, cats began to multiply and it became exceedingly harder for NASA to control them. This next picture suggests that the cats started having kittens in the fields surrounding the outdoor moon landing studios. The artificial moon surface became a huge playground for the cats. As they multiplied, it became literally impossible for NASA to get rid of them.

A Litter Of Kittens And A Prairie Dog Are Seen Playing On Moon Surface.

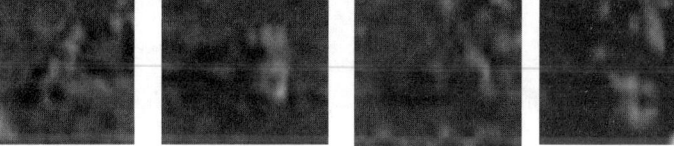

NASA ID# 10075955

Close ups

Prairie Dog

Kittens

 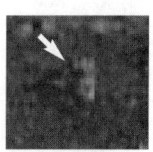 

 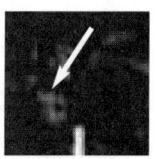

18-5

Chapter 18- The Moon's Animal Kingdom

In addition, dogs were found in many of the moon landing pictures; however they appeared to be pets rather than wild dogs. Below are a few of the moon landing photos containing dogs. If you were not looking for them, you may never have noticed the dogs because of NASA's day-to-night filters attached to their cameras.

Dogs Discovered On the Moon

Dog Sleeping

Same Dog Travels To A Different Location On the Moon.

It looks like the astronauts were disturbing the dog's nap and that is why it moved to a different location.

NASA ID# 20130756- Dog Sleeping

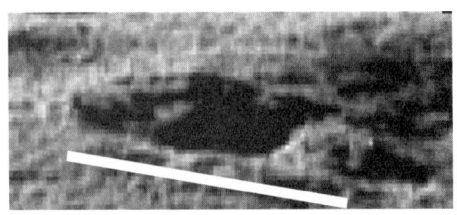

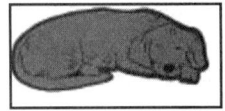

Dog Sleeping Under LM Spacecraft:

NASA ID# as11-40-5915

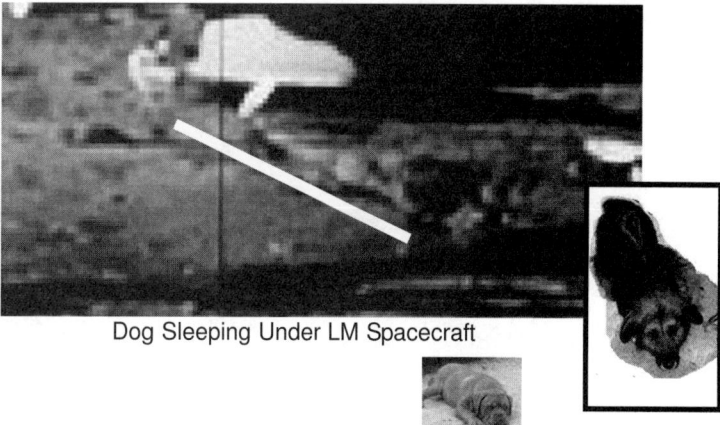

Dog Sleeping Under LM Spacecraft

Dog Sleeping On Hillside Of the Moon:

NASA ID # a15.1653138
Dog in laying in rock

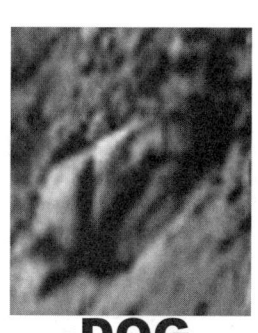

DOG

More Animals Found On Spaceship

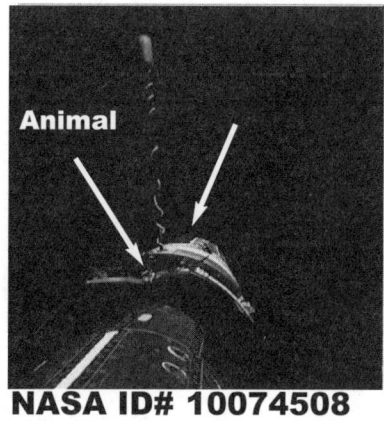

NASA ID# 10074508

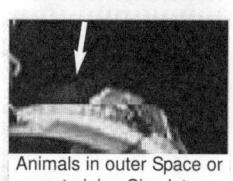

Animals in outer Space or on training Simulator

Chapter 18- The Moon's Animal Kingdom

Dangerous Desert Dragon

In addition to genital animals like dogs and cats, there were plenty of wild and dangerous animals that found their way onto the outdoor moon landing movie studios. The remaining animals illustrated in this chapter are the wild animals. These animals often posed a great danger to the astronauts.

NASA ID# 20121979-83
Animal, In Rock, Attacks Astronaut

Dangerous Moon Creature

Moon's first road kill

Look at how this Apollo astronaut accidentally ran over some kind of squirrel like creature on the Moon and killed it.

NASA ID# 20149543

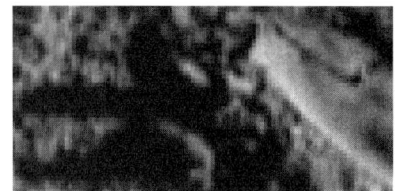

Maybe it wasn't an accident; look at the astronaut picking the creature up with his "Moon tongs". It is as though he intentionally hit it for dinner! One can just imagine how delicious squirrel would be compared to astronaut rations. "Jolly good shot mate."

Dead Squirrel for dinner as12-46-6932

Jolly good shot, mate!

Chapter 18- The Moon's Animal Kingdom

Snakes Found On the Moon

When considering the amount of snakes seen in NASA's Moon landing photos, evidence recommended that poisonous snakes were a big concern for everyone working at the outdoor moon landing studios. Below are a few illustrations of snakes found all over the place.

Snake Found Coiled-up In Rock

NASA ID# 20147668

Snake Discovered Coiled-up In Rock

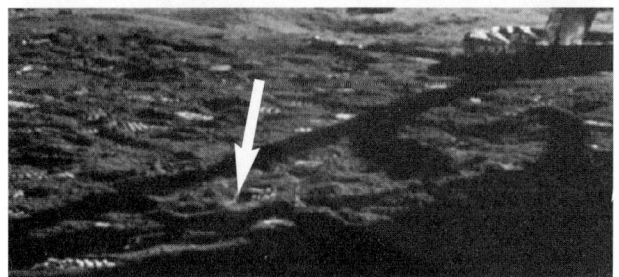

Very Large Snake
NASA ID# as11-10-5869

Snake NASA ID # as12-46-6933

18-9

NASA Tries Hiding Snakes in Moon Pictures.

Here is more proof that in many cases NASA artists simply painted over the animals that managed to get into the Moon landing photos. In this Apollo 17 photo a snake can be seen coiled up on the rock. As mentioned before snakes are very common in the Moon landing photos. The exciting thing about this snake picture is the matching next photo, which shows that this snake was painted over.

Snake on Rock
NASA ID # as17-145-22154

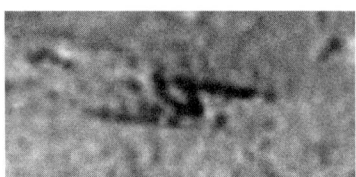

Close up of Snake

NASA s Artists Touch Up Photos and Cover up the Snake:

This picture is the touch up photo of the one above. The exciting thing about this picture is that while the artist was removing the snake he added a clue to the Moon landing hoax. The artist added a figurine depicting slave labor men pounding rocks. Was the artist trying to tell us he was a prisoner and NASA was using him as slave labor for the Apollo project?

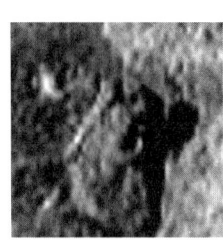

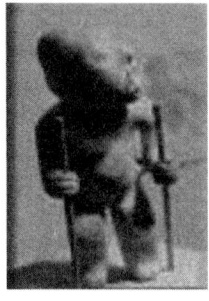

Chapter 18- The Moon's Animal Kingdom

Over-Night Visitors

There are many Moon landing photos containing animals that have gone undetected by NASA. These photos contain animals which managed to crawled onto NASA's outdoor studio sites and blended in so well that it was impossible to find every one of them. The next two sets of pictures are very good examples.

This Moon landing photography contains at least one white and black rabbit lying in black spray painted shadows and two small kittens.

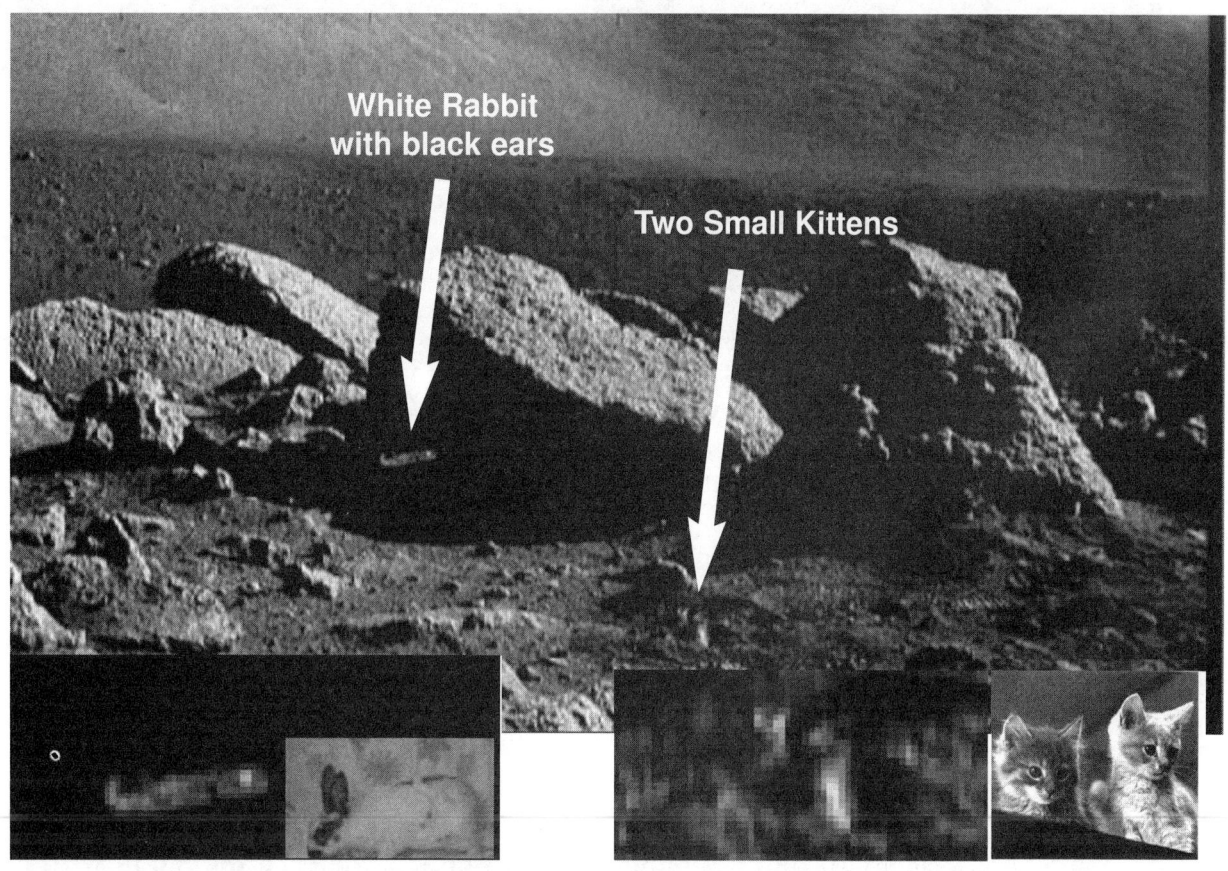

Birds Exposed On the Moon

It was also very common for birds to make their way into the moon landing portraits. Evidence suggests, using an airbrush technique, NASA artists often blended items like birds into the scene. This snapshot is an example. This is the original NASA picture and it is clear a bird that has landed on the equipment. There was also another picture of this same scene. However, this graphic had additional equipment added behind the bird and blended it right into the surroundings.

BIRD

NASA ID# 20135229

18-11

Chapter 18- The Moon's Animal Kingdom

Birds Hitch Ride To Moon On Spacecraft. These birds simply flew into the airport hanger where the Agena Target Docking Vehicle from the Gemini 8 Spacecraft space missions was being faked. Then the bird landed on the spacecraft and was detected by the photography editor. NASA then released this photo many years later in error.

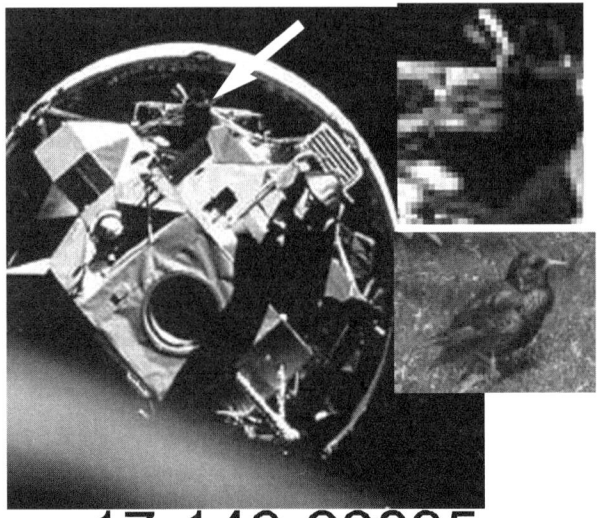

as17-148-22695

NASA ID# 10074212

A small Chickadee Bird

Owl and Mouse

An owl Swoops Down To Catch A Mouse That Hitched A Ride To the Moon With the Astronauts. This is no joke. Under a powerful magnified glass, it's clear that this is an owl trying to catch a mouse in outer space. Mice can can also be seen in other NASA photos including #20130572 and 20130713. Unfortunately, like many NASA moon landing pictures, they are simply too dark for paperbook format printing. To view these photos, you can go to either NASA's, or the Whiz-kid's web site. (Photo ID# as16-122-19533)

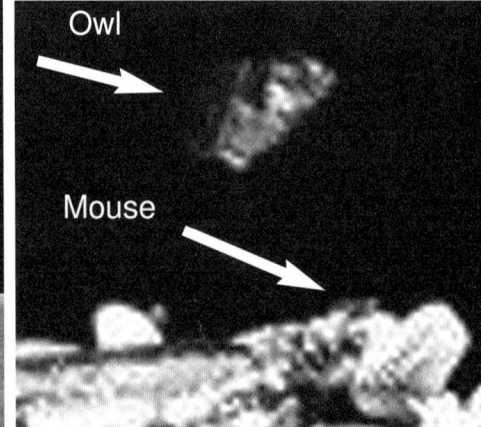

The mouse must have scaled across one wire supporting the spacecraft and was spotted by an owl up in the rafters of the simulator airport hanger.

Chapter 18- The Moon's Animal Kingdom

More Birds Revealed In the NASA Moon Landing Photography

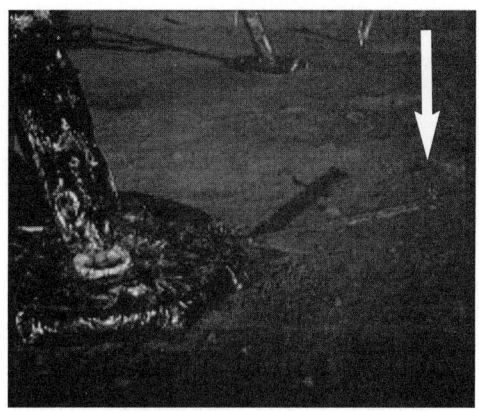

Bird

Bird Lands On Sensor Probe

Bird Reflection in Astronaut's Helmet

Bird Flying By in Moon Orbit or an Artist's Clue

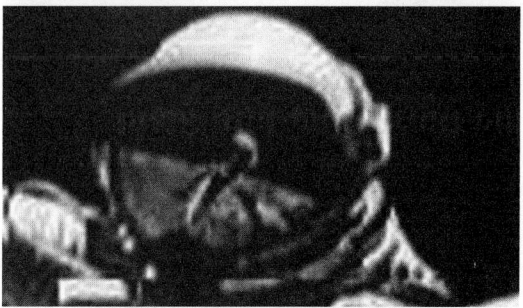

NASA ID# 10074018

18-13

Chapter 18- The Moon's Animal Kingdom

What about insects? It should have been impossible to keep insects off any outdoor movie studio. Undoubtedly, in the moon landing photos, insects were abundant and were simply edited out by NASA's team of airbrush artists. However, after carefully examining thousands of moon landing shots, many insects were discovered. Below are a few insect pictures that were discovered. Since we are talking about insects, a magnifying glass would be useful here.

Insects:

Black Ant Found On the Moon:

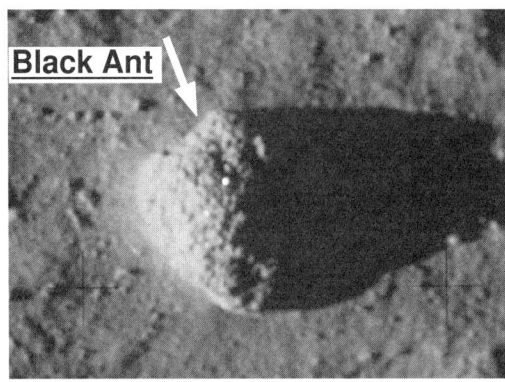
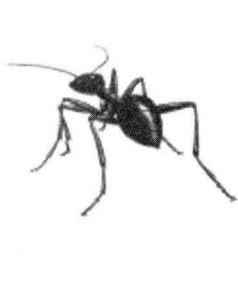

Caterpillars Uncovered:

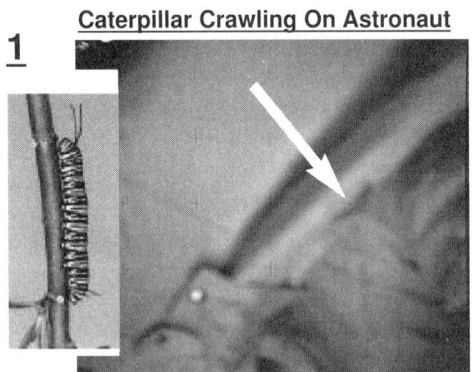

1

Caterpillar Crawling On Astronaut

NASA ID# 20130747

2

NASA ID# as11-40-5917

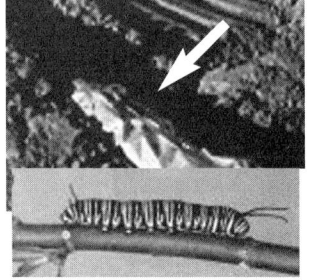

3

NASA ID# as12-46-6719

Butterflies On the Moon:

Butterfly on foorprint

Butterfly on Moon Rover

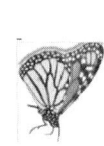

NASA ID# 20124033

NASA ID # 20130719

Butterfly

Chapter 18- The Moon's Animal Kingdom

With millions of different types of animals living on planet earth, it's impossible to determine the name of every animal in the moon landing portraits. Also, because many of them blend in so well to the artificial moon surface, it's often impossible to make out a complete picture of the animals. Below are samples of different animals discovered on the moon that have yet to be determined exactly what type of animal they are. If you would like to see these photos in a much higher image resolution, they are available at NASA.com or www.whizkids.tv.

Unrecognizable Creature Discovered On the Moon

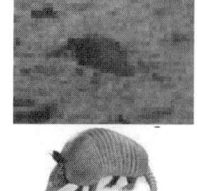

NASA ID# 20130779

Armadillo

NASA ID# 20121983 Small Animals

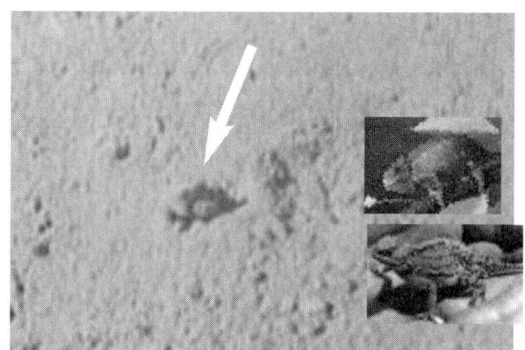

NASA ID# 20122165 Tiny Animals

NASA ID# 20147717 -Little Animals

NASA ID# 20134782 -Puppy

Possibly Armadillo Burrowing Into Sand

NASA ID# ap14-football

NASA ID# as11-40-5905

Evidence proving NASA was aware of these animals in the moon landing photography

Is there any evidence NASA was aware of these animals in the moon landing photography, and can that be proven? Well, the answer is YES, absolutely. Below are two groups of moon landing pictures containing animals. The pictures were taken at different locations, yet the NASA photo ID numbers are in sequential order. This reconfirms that NASA used a filing system for animals they discovered in their moon hoax pictures. In the below pictures, the animals have not been magnified, or pointed out. To maintain the original dimensions of these pictures, you would need a magnified glass to clearly see the animals.

Group 1 (10075620-10075630 All Contain Animals From Different Scenes.)

ID# 10075622

ID# 10075623

ID# 10075624

ID# 10075626

The Final Whizkids Experiment.

With so much evidence revealing the moon landings were fake, there is still the possibility that NASA officials may attempt to continue to deny the moon landings were fake, and that is not a good thing. However there is a simple solution to this problem, thanks to one of the Whizkids named Myrtle.

One afternoon while sitting with Myrtle and listening to her story about her family moonshine recipe, suddenly Myrtle came up with the perfect solution to prove government and NASA officials are not telling the truth about their claims of landing men on the moon. Myrtle was recalling some of the mobsters she knew about were running the country back in the 1920's and said "all those young men like Buzz, Neil, and the others should be forced to take a lie detector test, that will prove they are lying." Myrtle (age 92) probably came up with the simplest and most effective ways of proving NASA's Moon landings were fake.

CHAPTER 19

Advanced Theories

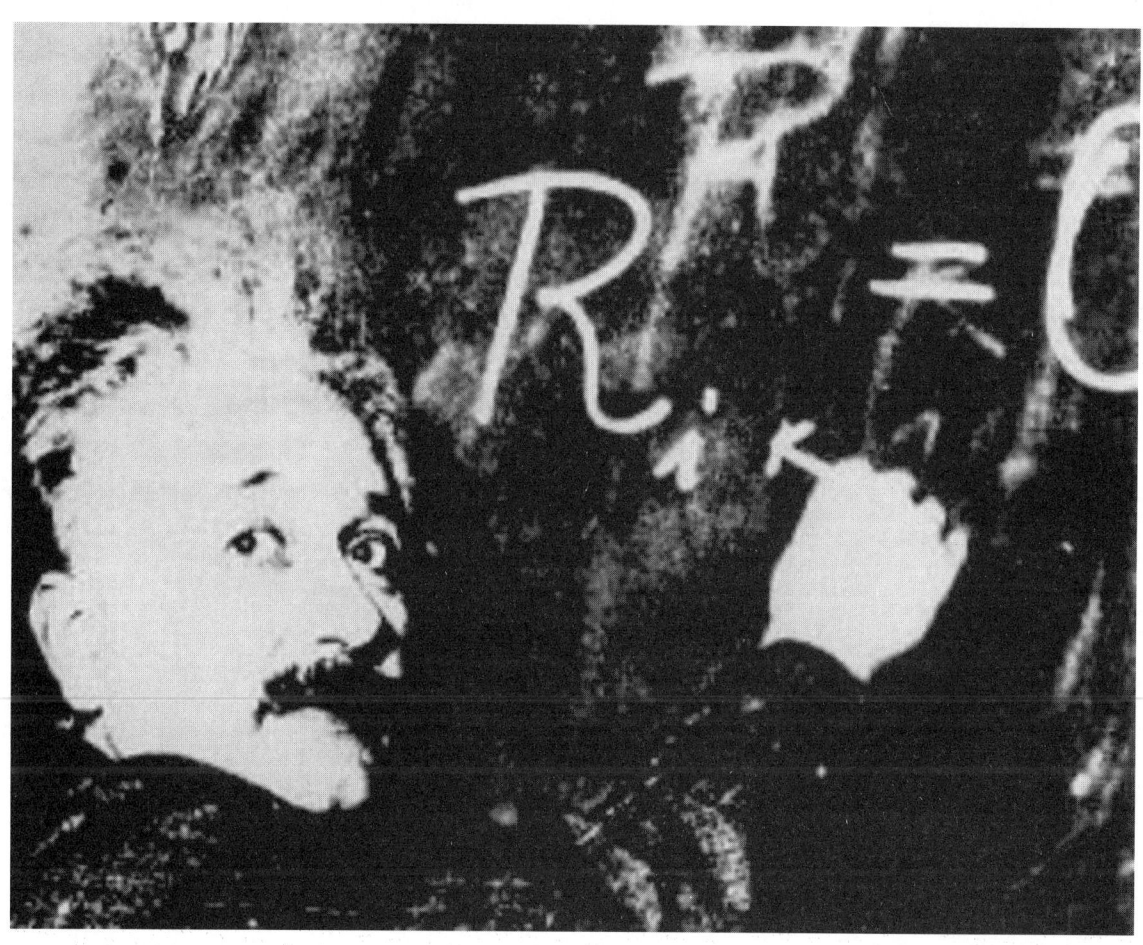

Chapter 19
Advanced Theories

Pictures, Supposedly Of the Earth Taken From the Moon, Were Totally Inaccurate And Out Of Phase

Here is a simple way to prove the photography taken from the Apollo missions were faked. During the time period of the Apollo mission, the phases of the earth to moon are totally incorrect in NASA's photos and clearly don't line up. These pictures must have been produced using the LOLA orbit simulator at Langley Research Center.

Back in 1969, everyone knew the moon went through phases, such as new moon, quarter-full, half-full, three-quarters-full, and full moon. People did not know that the earth, seen from the moon, has the same phases, and it takes just as long to complete the cycle. As the earth spins around once every 24 hours, the percentage of its illumination, seen from the moon, remains relatively close during any 24-hour period. Additionally, the phase of the moon appears not to change all that much from 8:00 p.m. on one night, to 8:00 p.m. on the next night. The percentage of the earth that appears lit remains constant during 24 hours.

NASA : AS11-44-6550f
Two thirds full earth phase

NASA : AS11-44-6642
Half full earth phase

NASA : AS11-44-6689
One thirds full earth phase

Knowing that the earth's illumination phases are identical to how we see the moon from the earth, we can easily prove that the Apollo 11 moon missions, by Astronauts Neil A. Armstrong and Edwin E. Aldrin Jr., were faked.

1) The picture on the left NASA claimed was taken on July 20, 1969. This was supposedly when man first walked on the moon. Notice that the earth phase is at two-thirds full.

2) The middle picture was supposedly taken on July 21, 1969. This was the next day. Note that the earth's phase mysteriously jumped to only one-half full.

3) NASA records claimed that the photo, on the far right, was taken on July 20, 1969. This was the same day as the first picture. The earth's phase has now jumped backwards to only one third of the full phase. NASA photo descriptions even confirmed this visual observation. (As photographed from the Apollo 11 Spacecraft, during its lunar landing mission, NASA's official photo description: one third of the earth's sphere illuminated earth's terminator, sunlight, and a portion of East Africa.)

These photos could not have been taken on the moon as NASA's claimed. There would have to be at least a 4 1/2 daytime frame between the earth's phases. During the Apollo 11 mission, the astronauts were not even on the moon 4 1/2 days to have taken portraits of the different earth phases.

The fact that the earth rotates under the sunlight much faster than the moon makes no difference one-way, or the other. Both the moon and earth complete their full illumination cycle in approximately 28 days. They change from new to full in about 14 days. They change from new to half-full in 7 days. In addition, they change from new moon to two thirds full in 9.3 days. Thus, the time it would take to change from one-half full to two-thirds full, is 2.3 days.

It also makes no difference whether the graphics were taken on the moon, or orbiting around the moon. At an altitude of 70 nautical miles (130 kilometers) above the moon's surface, the diameter of the moon is only 3,480 kilometers, yet the diameter of the earth is 12,753 kilometers. The lunar orbit was like being on the surface of a planet smaller than the earth, and we know the phase of the moon hardly changes at all from early in the evening to late in the morning.

So, why does NASA still show Apollo Mission pictures with total earth phase shifts taking place on the same day? It's because they dismantled the LOLA Orbiting Simulator and cannot replace the photos without being detected.

In the NASA images taken from the moon Earth's size in proportion to the Moon's surface should have remained constant, but It did not.

In proportion to the moon's surface, earth's size should have remained constant, but it did not in NASA's alleged moon landing photos. Once again proving NASA's Moon landing photos are fakes.

Here is a simple way to prove NASA's Apollo moon landing photos were faked. Since the distance between the earth to the moon remains relatively constant at all times, if you were to take a picture of the moon, it would remain the same size each and every night. Of course, it would go through its normal illumination cycles, but the actual diameter would remain the same. From the moon, the size of the earth also remains constant. From the moon surface, earth appears much larger because it is six times bigger than the moon. The only time the size of the earth would change is if you had a zoom magnification feature on your camera, which according to NASA's records, the astronauts hand held cameras did not.

Below are images taken from the moon in 1994 by the Clementine spacecraft, which shows the actual size of the earth from the moon. This should have been what was seen and filmed by the astronauts during their missions to the moon. However, NASA's photos and astronaut descriptions of the earth were totally incorrect, most of the time earth looked more like Venus. It obvious NASA at the time did not totally understand what the astronauts should have seen on the moon's surface.

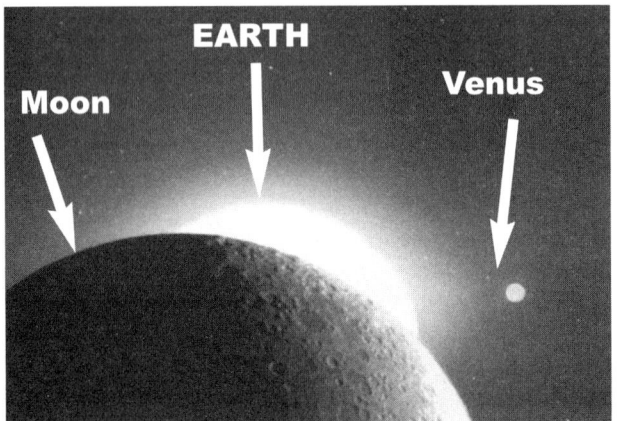

The large white ball on the far side of the moon is not the SUN it is Earth, reflecting about 30% of incident sunshine. The smaller bright light is the planet Venus that shines above as the solar corona peaks. And of course the hundreds of small lights are stars.

These next sets of NASA images clearly show there were several different sizes of the earth. The Apollo 11 picture, on the left, was the closest to what should have been seen by NASA astronauts. The picture on the far right, was the most unlikely example of what the earth should have looked like from the moon's surface. Yet NASA used it over and over again. Again it's totally out of proportion and more like the size of Venus not earth.

NASA ID# as11-44-6550f   apollo 16 mission   NASA # as17-134-20387

Using modern day scientific principles, NASA's pictures of earth don't look like what would be expected. The majority of NASA's photos were so imperfect that they made the earth look like a small marble.

The reason the smaller moon was used in the moon landing photos was because, at the time, when looking at the moon from earth, NASA scientists assumed the earth was the same size as the moon. Maybe they thought this because the earth and moon are the same distance apart. This was an easy mistake that anyone could have made.

Evidence supporting the theory that NASA scientists were responsible for the mistake of using a miniature sized earth was a statement made by Apollo 15 astronaut James Irwin. During an interview recorded on the film "For All Mankind", astronauts Irwin stated while standing on the moon, he could see the earth above them and "it looked about the size of a marble". Since we know the astronauts never went to the moon, they had know way of knowing what to say. They simply followed the script that was produced under the guidance of NASA's scientists, which indicated they believed that, from the moon's surface, the earth looked like a small marble or gulf ball, however that is just the opposite of what the astronauts would have seen.

NASA ID# as11-44-6550

Below are more official NASA images of earth, from the moon, that simply could not have taken place and are fakes. Besides the moon dimensions being incorrect these pictures also have other problems such as the astronauts helmet being open.

20117258

as17-134-20384

NASA #: as17-151-23188

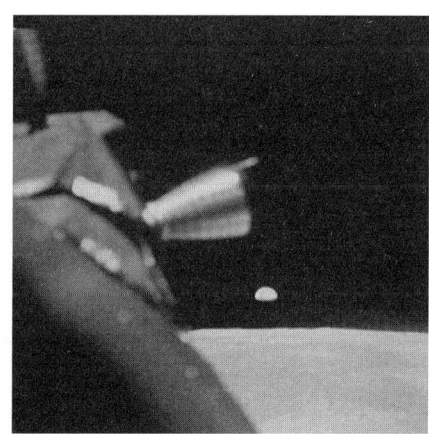

20130571

10075958

19-5

NASA's Own Evidence Proved the Moon's Transmissions Were Really From the Earth.

During the videos of the moon landings, the astronauts replied instantly to Mission Control, in Houston, with no time-lapse between audio signals. Yet, light, radio waves, and all energies of the electromagnetic spectrum travels at roughly 186,000 miles per second. The moon is approximately 240,000 miles from earth. This means that the response time of the astronauts to comments made by Mission Control should have been over two seconds. The filming must have taken place on earth.

NASA ID# 10075309

To make this easier, imagine it this way: Mission Control transmits a message to Apollo 11 on the moon surface saying, Neil and Buzz, can you get out of the LM and walk around (and remember to put your helmets on this time). To travel to the moon, this message takes just over a second. When Neil and Buzz receive it, they reply with "Yes boss".

Then, this message is transmitted all the way back to earth. It is received and broadcasted on the monitor in Mission Control. The astronaut's reply to NASA should have contained a time-lapse and never matched the video scene, but it did not. For this reason, the Astronauts had to be at an undisclosed location here on earth. There is clearly no other option. It simply cannot be a coincidence that everywhere you turn, something reveals a clue that the NASA moon landings were faked here on earth. For example, even this official NASA Apollo Mission Control photo revealed a clue that the NASA moon landings were faked here on earth.

One of NASA's employees intentionally pasted a monkey's head over President Nixon's head, and somehow for many years, it has been overlooked by NASA's watchdogs.

Monkeys Head On President Nixon

Astronaut's Moon Surface Vacuum Test Proves the Apollo Moon Landings Were Filmed Here On Earth.

Doubling the video speed achieved the appearance of earth-like gravity. If the Apollo video recording's speed was increased by a factor of two, this would have achieved the appearance of earth-like gravity. If the moon was one quarter of the earth's size, this would be expected, but it is not. The moon is one-sixth the size of the earth. To achieve earth-like gravity of actual moon footage, it would require increasing the speed by the square root of six. This is an approximate 2.45 x increase. Clearly, NASA scientists made a major miscalculation in the video speed differential. There is no way recordings could even be remotely mistaken as something filmed on the moon.

Feather fall next to stage prop levitation wire. NASA ID# 10100273

To prove the video recordings were not shot on the moon, one could observe the film footage from Apollo 15. In this footage, Astronaut David Scott demonstrated the Galilean principle by dropping a hammer and a feather to the ground, in a vacuum, on the moon. The decent to the ground followed a ballistic trajectory in accordance with Newton's laws of motion. The hammer was in freefall for at most 1.1 seconds.

Therefore, the distance it would have traveled is:
0.5 x [lunar] gravity x time x time =
0.5 x (9.8/6) x 1.1 x 1.1 = 0.99m.
Thus, the hammer could not have fallen a distance of more than 0.99m.
David Scott's height is clearly taller than Armstrong's (who is 5'11") by several inches. Scott was also wearing "Moon Boots". David Scott dropped the hammer from shoulder height, over 1.50m. This is not possible. Astronaut, David Scott, could not have been standing on the moon when he dropped the hammer.

However, the distance the hammer did travel does equal what would be expected if NASA did indeed film Astronaut David Scott on earth then doubled the speed.
0.5 x [earth] gravity x time x time =
0.5 x 9.8 x 0.55 x 0.55 = 1.48 m = 148 cm.

Again, the Galilean Experiment proved filming was not taken on the moon, but on earth! My calculations have been verified by several mathematicians, and feel free to verify them yourself.

The Communication Link NASA Claimed the Apollo Astronauts Used Could Not Function

Using a stationary satellite system attached to the Moon Rover Vehicle, the Astronauts were supposedly filmed many miles from the Lunar Lander Spacecraft. To see for yourself, look below. Live film broadcasts of the astronauts driving the Lunar Rover and making quick stops to pick up samples would not have been possible using this communication system. Stationary satellite configurations were often very difficult to calibrate. Furthermore, they must remain very still, and the slightest movement, or vibration, will cut off the communication link and stop the data transmission. To transmit the live video footage back to earth, the astronauts had to maintain a direct satellite link with earth. This could not have been possible because the astronauts were supposedly driving the Moon Rover.

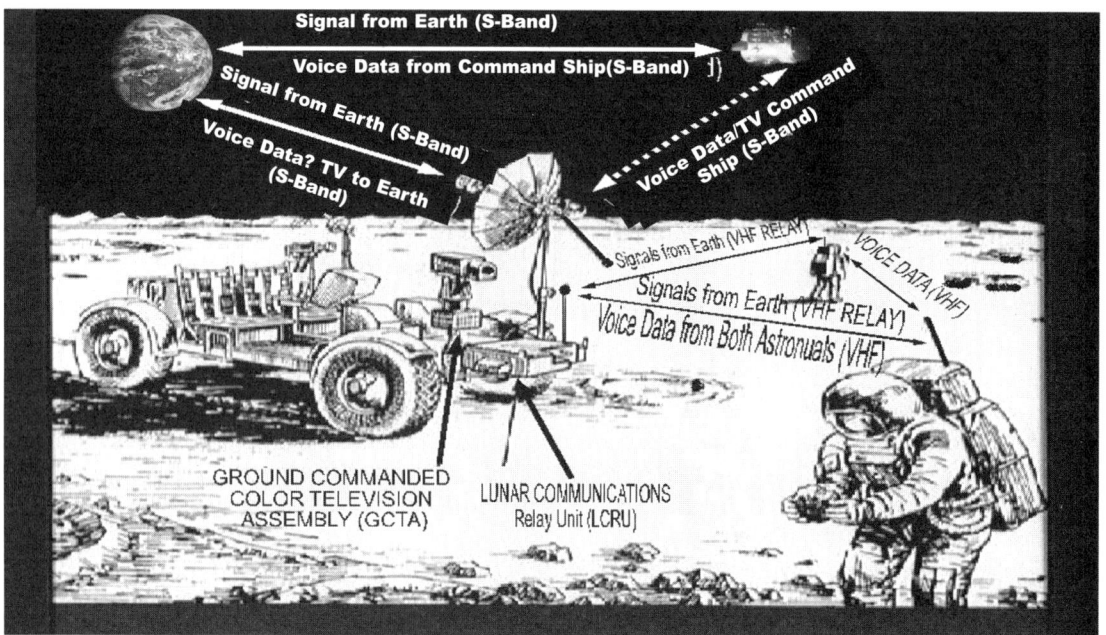

It could have easily taken the astronauts 30 minutes to establish a communications link each time the dish was moved. The calibration settings are that critical. The slightest vibration, or movement, would have disconnected their data link, yet the astronauts maintained a communication link while the satellite dish was bouncing all over the place. This is another perfect example of why NASA refuses to let the astronauts take any detailed questions about their moon missions. Everyone of the astronauts (actors) would have been required to know the calibration procedure of this satellite system. If questioned, none of the astronauts would have been able to explain how the system functioned.

There is no mystery to how NASA managed to make it look as if the astronauts' transmissions were coming from the moon. Everyone is just overlooking the intriguing fact that NASA launched the "TETR-A satellite" shortly before the alleged Apollo missions took place. The purpose of this special TETR-A satellite according to NASA was only to simulate transmissions coming from the moon in order for ground crews to practice a very realistic moon landing.

The question that needs to be answered is why couldn't NASA use this satellite to fake the transmission during the actual Apollo mission to the moon? Well NASA claimed this training simulator satellite mysteriously crashed shortly before the first Apollo mission, how convenient.

The obvious conclusion is NASA uses this satellite to relay the astronauts voice from earth, simulate instrument readings such as altitude, and telemetry data as if the transmissions were coming from an Apollo spacecraft during their mission hoax. Since less than a few hundred people even knew that this satellite existed and it's capabilities were easy enough for NASA to convince everyone the missions were real.

There was never any live broadcasts from the moon.

The television equipment that NASA claims the astronauts used on the moon was incapable of transmitting a live broadcast, because it functioned on slow band 10 frames per second (non-interlaced) and 320 lines per frame. In order to broadcast this type of signal to the waiting world, the pictures would have to first be converted to the commercial TV standards. In the US, the standard was the EIA (NTSC) standard of 30 frames per second (60 interlaced fields per second) at 525 lines per frame. In Australia, the standard was the higher resolution, CCIR 25 frames per second (50 interlaced fields per second) at 625 lines per frame.

Furthermore during the alleged Apollo 11 Mission, NASA claims a RCA scan-converter was used, which operated on an optical conversion principle. The video signal was displayed on a 10-inch black-and-white monitor and then a Vidicon TK22 camera pointed at the screen to record the image to be transmitting to the rest of the world.

If indeed this type of system was used as NASA claims, then these alleged live broadcasts from the moon could not have come from the astronaut's camera equipment. If a 10 frames per second live video feed was actually being displayed on the 10-inch monitor, then a Vidicon TK22 camera recording at 30 frames per second should have captured the blackened out frames between the much slower 10 frame video feed being displayed.

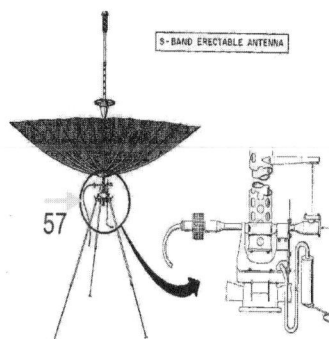

Furthermore, even if the Vidicon TK22 camera was gated to scan a single field at the EIA (NTSC) rate of 1/60th of a second, to not take a picture until the 10-inch monitor had completed displaying a full frame, the 10 television cathode ray tube would still need enough persistence to retain the picture, and RCA TV did not back then. To compensate for such loss of brightness between the top and bottom of the picture would require something like a 3 electrode Sony Trinitron 3 electrode gun tube technology, which was not developed until years later. Clearly the lack of available technology would also suggest NASA was incapable of ever transmitting a 30 frames per second (60 interlaced fields per second) live television broadcast from the moon, and their moon landings being a hoax.

NASA's Own Solar Wind Test Results Confirmed the Moon Landings Never Took Place

WhizKids Experiment Number 7

This experiment came from a group of college whiz kids at Harvard University. They would like to confirm the results of NASA's solar wind composition experiments taken during the Apollo 11, 12, 14, 15, and 16 missions. It consisted of an aluminum foil sheet, 1.4 meters by 0.3 meters, that was deployed on a pole facing the sun. On Apollo 16, a platinum sheet was also used. This foil was exposed to the sun for periods ranging from 77 minutes on Apollo 11 to 45 hours on Apollo 16. This allowed solar wind particles to embed themselves into the foil. The foil was then returned to earth for laboratory analysis.

The Harvard University WhizKids would like to test the chemical composition of the embedded solar wind and measure isotopes of the light noble gases, including helium-3, helium-4, neon-20, neon-21, neon-22, and argon-36. After that, they would like to test the variation in the composition in correlation with variations in the intensity of the solar wind. They would determine this from magnetic field measurements. Next, the Whiz kids would observe the measurements from different missions. Extensive chemical composition testing of the solar wind particles would certainly confirm the validly of the Apollo missions.

Chapter 20

Whistle-Blower Clues

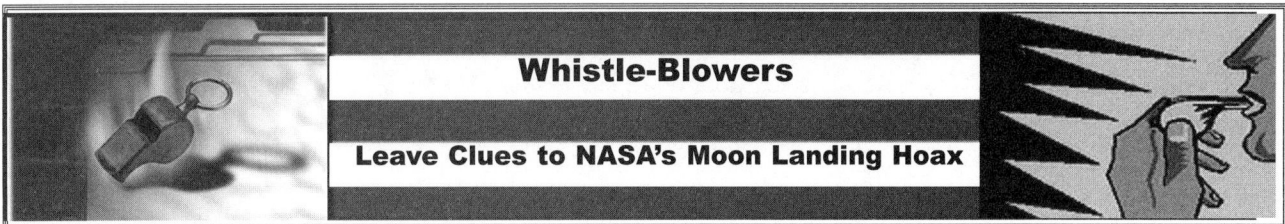

Chapter 20
Whistle Blower Clues

With the overwhelming number of intentional clues discovered in NASA's alleged Moon Landing evidence of proof, why haven't the people who left these clues publicly expressed their disapproval of the fake moon landings?

Even NASA tries to convince people their moon landings must have been real because there has been plenty of time for someone to come forward and admit their involvement in a hoax, yet no one has done so.

For NASA to ask people to consider the fact that no one has ever confessed as proof the moon landings were real indicates NASA is these confession up. The reality is more likely than not, many people working on NASA Apollo moon landing hoax have already left confessions based on the number of whistle blower clues already discovered in NASA's moon landing records. NASA and the CIA are simply preventing these incriminating confessions from reaching the public.

Can you imagine someone approaching one of the Television news networks all excited about a written deathbed confession their Uncle Wernher von Braun gave them about the moon landings being faked? Initially one might think the news networks would be all excited about the letter and stop all the programming running for an important news flash. However what is more likely going to happen is; the people with the deathbed confession letter are going to be escorted out of the building by the News station's security. Then the news station is going to immediately contact the CIA to have these people picked up and their uncle Wernher deathbed confession letter confiscated.

Although news stations are permitted to air stuff that is speculative in nature about the moon landings being fake, they are certainly not permitted to air any solid evidence that would prove conclusively the American Moon Landings were fake. We confirmed this after approached several national News Networks about the film footage we found at the local library that contains earth creatures moving around in the Apollo Mission action scene. At first the news station thought we were crazy and said no station would air such a ridiculous thing. However after directing them to the moonbloopers.com web page for a sneak preview of a stream video, their attitude quickly turned very negative.

Two of the major news networks we contacted claimed it would be impossible for them to air this information under the Espionage Law, which governs information that could affect the national defense of the United States. One network representative claimed she had seen this type of moon landing hoax evidence before and we must immediately stop distributing it or the CIA would certainly be paying us a visit soon. We asked her, doesn't the First Amendment of the United States constitution "freedom of speech" give you the right to air our story? She started laughing and said, "You have got to be joking. The United States television news media is one of the most government-regulated industries ever imaginable. We can't make a move without worrying what topics and words can or cannot be covered. We are not allowed to show many things that could potentially have a big effect on the world and to the American people. I was just forwarded a memo that the government is banning us from showing any injured soldiers returning from the Iraq war or to the funerals, which is a good thing, but it never ends. The government regulates almost everything and your evidence showing the moon landings were faked would be considered the top of the list. She then claimed it was NASA that even started to make the 7-second delay a requirement for television broadcasting during the Apollo Missions. Before the Apollo moon landings broadcast there was no system in place to hide stuff from the viewers and live broadcast was really live. We took this nice old lady's advise and determined it was not in our best interest to continue broadcasting the videos over the Internet. We would wait until the completion our research and more solid evidence was also obtained confirming NASA faked the moon landings.

Is it NASA who is leaving these clues to the moon landing being faked?

Based on our initial encounter with the news media it was clear it could take years before something like our video would be permitted to air on a National News Broadcast. We saw such a fear instilled into the National News Networks by the Untied States' CIA there is the strong possibility NASA has been regulating and possibly even producing all the conspiracy broadcast up until now.

Chapter 20 - Whistle Blower Clues

Some of these clues of the moon landings are so obvious, so why did NASA has released them? Could NASA be so careless with their evidence to release hundreds of whistle blower clues undetected, this would seem unlikely. Whether the Apollo Moon Landing Missions were real or not. One thing is for certain, NASA is not stupid, they employ many of the best brains in the World. They should be have been able to easily spot some of the most obvious Whistle Blowers clues found in the moon landing photography before releasing them. In addition the United States' Military has a history of proving they are masters at keeping projects like this a secret. The development of the nuclear bomb would be a perfect example how they kept over ten thousand people silenced about the project. According to the people working on the project they never knew that they were building nuclear bombs. Because NASA built these huge secret cities where they worked and everything was divided into different development sections. They were instructed never to speak about what they were working on to anyone else working in the different areas of the secret city or tell their immediate family members about what they were doing. If anyone was caught speaking to someone else about their work they could be imprisoned along with the people they had spoken to. Other than Albert Einstein most of them claimed to be totally unaware of what they were working on until after the bombs were used on Japan, only then did they know what was going on.

The point here is it is simply not possible for anyone working on the moon hoax project to make a confession and have it reach the news media. Can you imagine the Apollo Astronaut Neil Armstrong showing up at the Fox News Headquarters and saying, "I have a whopper of a story to tell you, we faked the moon landings!

The conclusion to a few of the more obvious clues revealing the moon landing were fake is that there were released intentionally by NASA in order to break the news to the world gradually. And the majority of the whistle blower clues presented in this book were simply difficult to recognize or see because no one really knew what to be looking for, and the photos were released as ok. To help prove this point the remainder of this chapter contains several alleged Moon landing photos that contain whistle blower clues proving the photos were faked.

Imagine NASA calls you up and offers you the dream job which pays ten million dollars ($10,000,000.00) a year with a five million dollar ($5,000,000.00) sign-up bonus. Here is the catch, you'll only be able to work for one year and it a top secret job, you can never tell anyone one for the rest of your life what you worked on. They then tell you the Apollo Moon landings were a hoax and they want you to look for defects in the photos that would reveal to the world, they were faked. Would you take the job?

Congratulation on your new job! Lets test your skill and see how good you do on your first day on the job. The rest of this chapter is filled with photos not yet mentioned in this book and this chapter also contains certain items that reveal these NASA moon-landing photos are fakes. Take a few minutes and see if you can detect the problems with these NASA photos and if they would have slipped by you undetected.

For example this next picture was illustrated in the book earlier about it problems and helps illustrate what types of things you would be expected to find in the remaining photos. This NASA photo has the marking A Dog in the picture. It also has the letter A and a drawing of a dog. This lead to the discovery of a dog in the moon landing picture. See if you can determine what is wrong with the rest of the NASA Moon landing images on the next few pages.

Example of what to look for in testing your skill level as a whistle blower clue screener for NASA.

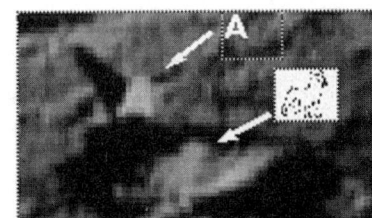

NASA ID# 20130804 dog and letter A

Chapter 20 - Whistle Blower Clues

Here you can test your skills to see if you would have been a good Whistle Blower Clue Screener for NASA. These images all contain items which reveal they were not taken on the moon, how many can you find?

NASA ID# as16-113-18342

_____
_____

NASA ID# AS16-0700

NASA ID# as11-40-5943

_____
_____

NASA ID# as11-40-5920

NASA ID# 20147850

_____
_____

20-4

Chapter 20 - Whistle Blower Clues

NASA ID# 10100137

NASA ID# 10075624

NASA ID# 20121966

LZ

Special Memorabilia Whistle Blower Clue left by the famous Greek God Himself Apollo!

NASA ID #: as12-46-6822 h

Meet Apollo the Greek mythology God, son of Zeus and Leto

Chapter 20 - Whistle Blower Clues

10075310

10100268

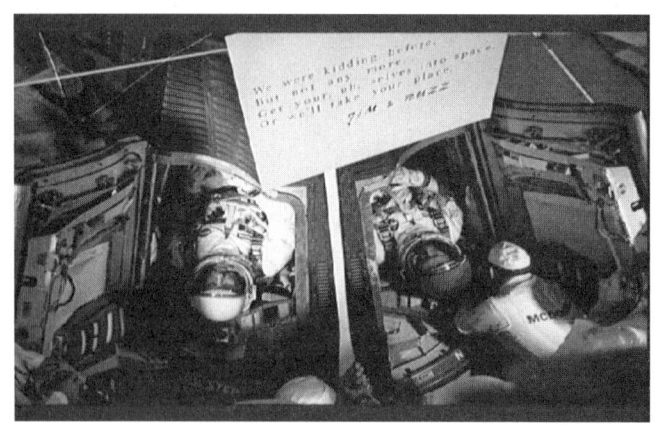
NASA ID# 10074365

Chapter 20 - Whistle Blower Clues

as12-46-6806

20122442

a16v1235056

a16v1235054

_____
_____

NASA ID# as11-40-5931HR
Hint: Follow the yellow brick road

NASA ID# 10100095

Chapter 20 - Whistle Blower Clues

All of these photos had the word like "FAKE" written into the artificial moon surface, see if you can find the words.

AS15-82-11117

AS15-82-11-11169

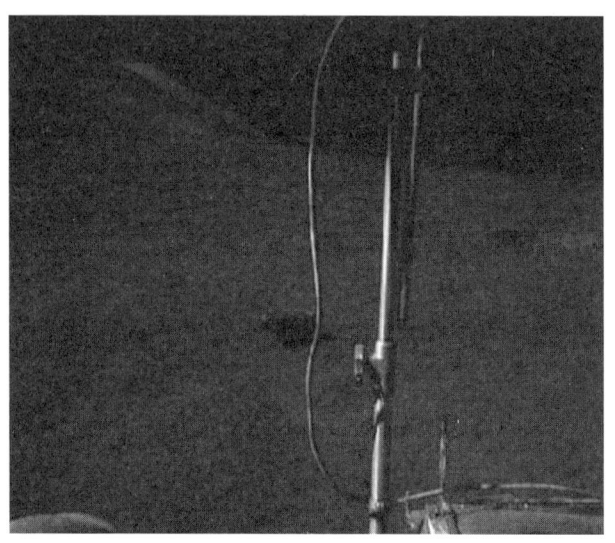

NASA ID # AS15/82/11142

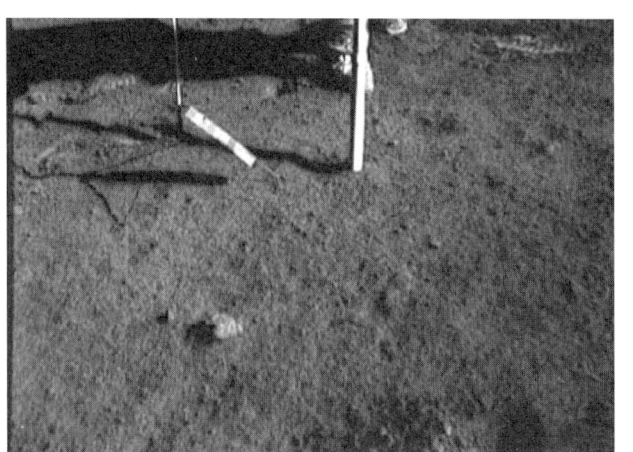

20-8

Space Science Glossary

Glossary

A

Accretion - Accumulation of dust and gas into larger bodies such as stars, planets and moons

Accretion disk - A relatively flat sheet of gas and dust surrounding a newborn star, a black hole, or any massive object growing in size by attracting material

Actinide -any of a series of chemically similar, mostly synthetic, radioactive elements with atomic numbers ranging above 89
Active galactic nuclei (AGN) It is believed that these are normal galaxies with a massive black hole accreting gas at its center, thus producing enormous amounts of energy at all wavelengths of the electromagnetic spectrum

Angular size - width or diameter of an object measured as an angle from the observer's eyes, a way of stating the diameter/distance ratio for an object

Angstrom- A unit of length equal to 0.00000001 centimeters. Scientists sometimes write this as 1×10^{-8} cm (see scientific notation)

Apoapsis- The point in an orbit when the two objects are farthest apart. Special names are given to this orbital point for commonly used systems. For example, the point of greatest separation of two stars, as in a binary star orbit, is called apastron; the point in its orbit where a planet is farthest from the Sun is called aphelion; the point in its orbit where an Earth satellite is farthest from the Earth is called apogee

Apparent brightness- brightness of an object as seen by an observer dependent upon the object's wattage, if it is self-emitting, or upon the amount of light scattered or reflected

Apparent magnitude - a measure of the apparent brightness of an astronomical object, as observed from Earth. The scale of magnitudes is logarithmic.

Asteroid- Chunks of rock that travel through space. The biggest asteroids are hundreds of miles wide, but most are as small as pebbles. Most asteroids lie in a ring, or belt, around the sun between the planets Jupiter and Mars.

Astronaut - a person who travels in a spacecraft, esp. a crew member

Astronomical unit (a.u.) -a unit of distance equal to the average distance between the earth and the sun, about 93 million miles; 149,597,870 km

Astronomy -The scientific study of matter in outer space, especially the positions, dimensions, distribution, motion, composition, energy, and evolution of celestial bodies and phenomena

Astrophysics- The part of astronomy that deals principally with the physics of stars, stellar systems, and interstellar material

Atmosphere- the gaseous mass or envelope surrounding a celestial body in space; The gas that surrounds a planet or star. The Earth'satmosphere is made up of mostly nitrogen, while the Sun's atmosphere consists of mostly hydrogen

Atom- a unit of matter, the smallest unit of an element, consisting of a dense, positively charged nucleus surrounded by a system of negatively charged electrons

Atomic- 1) of or relating to an atom, 2) of or employing nuclear energy, 3) very small

B

Binary stars- Binary stars are two stars that orbit around a common center of mass. An X-ray binary is a special case where one of the stars is a collapsed object such as a white dwarf, neutron star, or black hole. Matter is stripped from the normal star and falls onto the collapsed star, producing X-rays

Black hole- An object whose gravity is so strong that not even light can escape from it

Blackbody radiation- The radiation or the radiance at particular frequencies all across the spectrum produced by a blackbody ,that is, a perfect radiator (and absorber) of heat. Physicists had difficulty explaining it until Planck introduced his quantum of action

Blackbody temperature- The temperature of an object if it is re-radiating all the thermal energy that has been added to it; if an object is not a blackbody radiator, it will not re-radiate all the excess heat and the leftover will go toward increasing its temperature

Bremsstrahlung- "braking radiation", the main way very fast charged particles lose energy when traveling through matter. Radiation is emitted when charged particles are accelerated. In this case, the acceleration is caused by the electromagnetic fields of the atomic nuclei of the medium.

C

Calibration- A process for translating the signals produced by a measuring instrument (such as a telescope) into something that is scientifically useful. This procedure removes most of the errors caused by environmental and instrumental instabilities.

Cluster of galaxies - A system of galaxies containing from a few to a few thousand member galaxies which are all gravitationally bound to each other.

Collecting area - The amount of area a telescope has that is capable of collecting electromagnetic radiation. Collecting area is important for a telescope's sensitivity: the more radiation it can collect (that is, the larger its collecting area), the more sensitive it is to dim objects.

Comet- Comets are made of dust and ice and look like dirty snowballs. There are millions of them traveling through space.

Constellation- A group of stars that makes a pattern. There are 88 constellations in our sky.

Constellations- seem to twinkle in the sky. This happens because moving air blurs starlight as it travels to Earth.

Corona (plural: coronae)- The uppermost level of the solar atmosphere, characterized by low densities and high temperatures (> 1,000,000 degrees K).

Cosmic background radiation primal glow - The background of radiation mostly in the frequency range 3×10^8 to 3×10^{11} Hz (see scientific notation) discovered in space in 1965. It is believed to be the cosmologically redshifted radiation released by the Big Bang itself.

Cosmic rays- Atomic nuclei (mostly protons) and electrons that are observed to strike the Earth's atmosphere with exceedingly high energies.

Cosmological distance- A distance far beyond the boundaries of our Galaxy. When viewing objects at cosmological distances, the curved nature of spacetime could become apparent. Possible cosmological effects include time dilation and redshift.

Cosmology- The astrophysical study of the history, structure, and constituent dynamics of the universe.

D

Declination - A coordinate which, along with Right Ascension, may be used to locate any position in the sky. Declination is analogous to latitude for locating positions on the Earth.

Deconvolution - An image processing technique that removes features in an image that are caused by the telescope itself rather than from actual light coming from the sky.

Density - The amount of mass of any substance which can be contained in one cubic centimeter. Measured in grams per cubic centimeter (or kilograms per liter); the density of water is 1.0; iron is 7.9; lead is 11.3.

Disk (of planet or other object) - The apparent circular shape that the Sun, a planet, or a moon displays when seen in the sky or through a telescope.

E

Eccentric - on-circular; elliptical (applied to an orbit).

Eccentricity- A value that defines the shape of an ellipse or planetary orbit. The eccentricity of an ellipse (planetary orbit) is the ratio of the distance between the foci and the major axis.

Space Science Glossary

Eclipse- The cutting off, or blocking, of light from one celestial body by another.

Ecliptic- The plane of Earth's orbit about the Sun

Ejecta - Material that is ejected. Used mostly to describe the content of a massive star that is propelled outward in a supernova explosion.

Electromagnetic spectrum - The full range of frequencies, from radio waves to gamma-rays, that characterizes light.

Electromagnetic waves- (radiation) - Another term for light. Light waves are fluctuations of electric and magnetic fields in space.

Electron- A particle commonly found in the outer layers of atoms with a negative charge. The electron has only 0.0005 the mass of the proton.

Electron volt- The change of potential energy experienced by an electron moving from a place where the potential has a value of V to a place where it has a value of (V+1 volt). This is a convenient energy unit when dealing with the motions of electrons and ions in electric fields; the unit is also the one used to describe the energy of X-rays and gamma-rays. A keV (or kiloelectron volt) is equal to 1000 electron volts. An MeV is equal to one million electron volts. A GeV is equal to one billion (10^9) electron volts. A TeV is equal to a million million (10^{12}) electron volts.

Elements- The fundamental kinds of atoms that make up the building blocks of matter, which are each shown on the periodic table of the elements. The most abundant elements in the universe are hydrogen and helium. These two elements make up about 80 and 20 % of all the matter in the universe respectively. Despite comprising only a very small fraction in the universe, the remaining heavy elements can greatly influence astronomical phenomena. About 2 % of the Milky Way's disk is comprised of heavy elements.

Ellipse - Oval. That the orbits of the planets are ellipses, not circles, was first discovered by Johannes Kepler based on the careful observations by Tycho Brahe.

Event horizon- The radius that a spherical mass must be compressed to in order to transform it into a black hole, or the radius at which time and space switch responsibilities. Once inside the event horizon, it is fundamentally impossible to escape to the outside. Furthermore, nothing can prevent a particle from hitting the singularity in a very short amount of proper time once it has entered the horizon. In this sense, the event horizon is a "point of no return".

Evolved star- A star near the end of its lifetime when most of its fuel has been used up. This period of the star's life is characterized by loss of mass from its surface in the form of a stellar wind.

Expanding Universe- Astronomers have discovered that distant galaxies are moving away from the Milky Way and also from each other. The whole universe is expanding or becoming bigger.

Extragalactic - Outside of, or beyond, our own galaxy.

F

False Color- Color added to a photograph of an object, such as a galaxy, to make it look clearer.

Frequency - A property of a wave that describes how many wave patterns or cycles pass by in a period of time. Frequency is often measured in Hertz (Hz), where a wave with a frequency of 1 Hz will pass by at 1 cycle per second.

G

Galactic halo - A spherical region surrounding the center of a galaxy. This region may extend beyond the luminous boundaries of the galaxy and contain a significant fraction of the galaxy's mass. Compared to cosmological distances, objects in the halo of our galaxy would be very nearby.

Galaxy - A component of our universe made up of gas and a large number (usually more than a million) of stars held together by gravity.

Galilei, Galileo (1564 - 1642) - An Italian scientist, Galileo was renowned for his epoch making contribution to physics, astronomy, and scientific philosophy. He is regarded as the chief founder of modern science. He developed the telescope, with which he found craters on the Moon and discovered the largest moons of Jupiter. Galileo was condemned by the Catholic Church for his view of the cosmos based on the theory of Copernicus.

Gamma-ray- The highest energy, shortest wavelength electromagnetic radiations. Usually, they are thought of as any photons having energies greater than about 100 keV.

Gravitational collapse- When a massive body collapses under its own weight. (For example, interstellar clouds collapse to become stars until the onset of nuclear fusion stops the collapse.)

Gamma-Ray Burst (GRB) Plural is GRBs.- A burst of gamma-rays from space lasting from a fraction of a second to many minutes. There is no clear scientific consensus as to their cause or even their distance.

General relativity- The geometric theory of gravitation developed by Albert Einstein, incorporating and extending the theory of special relativity to accelerated frames of reference and introducing the principle that gravitational and inertial forces are equivalent.

Giant Molecular Cloud (GMC)- Massive clouds of gas in interstellar space composed primarily of hydrogen molecules (two hydrogen atoms bound together), though also containing other molecules observable by radio telescopes. These clouds can contain enough mass to make several million stars like our Sun and are often the sites of star formation.

Gravity- A mutual physical force between two bodies. Most agree that it is an attracting force while some think that it may be a force that pushes the bodies together.

Guest star- The ancient Chinese term for a star that newly appears in the night sky, and then later disappears. Later, the Europeans called this a nova.

H

I

Implosion- A violent inward collapse. An inward explosion.

Infrared- Electromagnetic radiation at wavelengths longer than the red end of visible light and shorter than microwaves (roughly between 1 and 100 microns). Almost none of the infrared portion of the electromagnetic spectrum can reach the surface of the Earth, although some portions can be observed by high-altitude aircraft (such as the Kuiper Observatory) or telescopes on high mountaintops (such as the peak of Mauna Loa in Hawaii).

Inclination- The inclination of a planet's orbit is the angle between the plane of its orbit and the ecliptic; the inclination of a moon's orbit is the angle between the plane of its orbit and the plane of its primary's equator.

Image- n astronomy, a picture (photo) of the sky.

Interstellar medium- The gas and dust between stars, which fills the plane of the Galaxy much like air fills the world we live in. For centuries, scientists believed that the space between the stars was empty. It wasn't until the eighteenth century, when William Herschel observed nebulous patches of sky through his telescope, that serious consideration was given to the notion that interstellar space was something to study. It was only in the last century that observations of interstellar material suggested that it was not even uniformly distributed through space, but that it had a unique structure.

Iions- An atom with one or more electrons stripped off, giving it a net positive charge.

Ionic - (or ionized) gas Gas whose atoms have lost or gained electrons, causing them to be electrically charged. In astronomy, this term is most often used to describe the gas around hot stars where the high temperature causes atoms to lose electrons.

J

Jets - Beams of particles, usually coming from an active galactic nucleus or a pulsar. Unlike a jet airplane, when the stream of gas is in one direction, astrophysical jets come in pairs with each jet aiming in opposite directions.

K

Kelvin (after Lord Kelvin, 1824 - 1907) - A temperature scale often used in sciences such as astronomy. The fundamental SI unit of thermodynamic temperature defined as 1/273.16 of the thermodynamic temperature of the triple point of water. The Kelvin temperature scale is just like the Celsius scale except that the freezing point of water, zero degrees Celsius, is equal to 273 degrees Kelvin. (K = C + 273o) (F = 9/5C + 32o)

Kilogram (kg) - One kilogram is equivalent to 1,000 grams or 2.2 pounds; the mass of a liter of water. The fundamental SI unit of mass, it is the only SI unit still maintained by a physical artifact: a platinum-iridium bar kept in the International Bureau of Weights and Measures at Sevres, France.

Kinematics- Refers to the calculation or description of the underlying mechanics of motion of an astronomical object. For example, in radioastronomy, spectral line graphs are used to determine the kinematics or relative motions of material at the center of a galaxy or surrounding a star as it is born.

L

Light- Electromagnetic radiation that is visible to the human eye.
Light curve- A graph that displays the time variation in light or magnitude of a variable or eclipsing star.
Light year- A unit of length used in astronomy which equals the distance light travels in a year. At the rate of 300,000 kilometers per second (671 million miles per hour), 1 light-year is equivalent to 9.46053×10^{12} km, 5,880,000,000,000 miles or 63,240 a.u.
Limb- The outer edge of the apparent disk of a celestial body.

M

Magnetic field- A condition found in the region around a magnet or an electric current, characterized by the existence of a detectable magnetic force at every point in the region.

Magnetic pole- Either of two limited regions in a magnet at which the magnet's field is most intense.

Magnetosphere- The region of space in which the magnetic field of an object (e.g., a star or planet) dominates the radiation pressure of the stellar wind to which it is exposed.

Magnetotail- The portion of a planetary magnetosphere which is pushed in the direction of the solar wind.

Magnitude- The degree of brightness of a celestial body designated on a numerical scale, on which the brightest star has magnitude -1.4 and the faintest visible star has magnitude 6, with the scale rule such that a decrease of one unit represents an increase in apparent brightness by a factor of 2.512; also called apparent magnitude.

Mass- A measure of the total amount of material in a body, defined either by the inertial properties of the body or by its gravitational influence on other bodies.
Matter- A word used for any kind of stuff which contains mass.

Mega-ton- A unit of energy used to describe nuclear warheads. The same amount energy as 1 million tons of TNT. 1 megaton = 4×10^{16} ergs = 4×10^{9} joules.

Meteor or shooting stars- Small rocks and dust left behind from comets.

Meteorite- Chunks of rock that don't burn up when they enter the atmosphere and fall to earth.

Meter m - The fundamental SI unit of length, defined as the length of the path traveled by light in vacuum during a period of 1/299 792458 s. A unit of length equal to about 39 inches. A kilometer is equal to 1000 meters.

Microwave- Electromagnetic radiation which has a long wavelength (between 1 mm and 30 cm). Microwaves can be used to study the Universe, communicate with satellites in Earth orbit, and cook popcorn.

Milky Way- A spiral shaped galaxy in which the Earth is located. The sun and the earth lie in one of the spiral arms near the edge of the galaxy.

Moon- A natural object in space that travels around a planet. A moon is smaller than its planet.

N

Nebula- A diffuse mass of interstellar dust and gas inside a galaxy.

Neutrino- A fundamental particle produced in massive numbers by the nuclear reactions in stars; they are very hard to detect because the vast majority of them pass completely through the Earth without interacting.

Neutron- A particle commonly found in the nucleus of atoms with approximately the mass of a proton, but zero charge.

Neutron star- The imploded core of a massive star produced by a supernova explosion. (typical mass of 1.4 times the mass of the Sun, radius of about 5 miles, density of a neutron.) According to astronomer and author Frank Shu, "A sugar cube of neutron-star stuff on Earth would weigh as much as all of humanity!" Neutron stars can be observed as pulsars.

Newton's- law of universal gravitation (Sir I. Newton) - Two bodies attract each other with equal and opposite forces; the magnitude of this force is proportional to the product of the two masses and is also proportional to the inverse square of the distance between the centers of mass of the two bodies.

Noise- The random fluctuations that are always associated with a measurement that is repeated many times over. Noise appears in astronomical images as fluctuations in the image background. These fluctuations do not represent any real sources of light in the sky, but rather are caused by the imperfections of the telescope. If the noise is too high, it may obscure the dimmest objects within the field of view.

Nova (plural: novae) - A star that experiences a sudden outburst of radiant energy, temporarily increasing its luminosity by hundreds to thousands of times before fading back to its original luminosity.

Nuclear fusion- nuclear process whereby several small nuclei are combined to make a larger one whose mass is slightly smaller than the sum of the small ones. The difference in mass is converted to energy by Einstein's famous equivalence "Energy = Mass times the Speed of Light squared". This is the source of the Sun's energy.

O

Occultation - The blockage of light by the intervention of another object; a planet can occult (block) the light from a distant star.

Opacity - A property of matter that prevents light from passing through it; non-transparent. The opacity or opaqueness of something depends on the frequency of the light. For instance, the atmosphere of Venus is transparent to ultraviolet light, but is opaque to visual light.

Orbit- The path of an object that is moving around a second object or point.

P

Pair production- The physical process whereby a gamma-ray photon, usually through an interaction with the electromagnetic field of a nucleus, produces an electron and an anti-electron (positron). The original photon no longer exists, its energy having gone to the two resulting particles. The inverse process, pair annihilation, creates two gamma-ray photons from the mutual destruction of an electron/positron pair.

Parallax - The angle between the two straight lines that join a celestial body to two different points of observation; e.g., two different points on the Earth as it moves through space.

Parsec - A large distance often used in astronomy, it is equal to 3.26 light years, or 3.1×10^{18} cm (see scientific notation). A kiloparsec (kpc) is equal to 1000 parsecs. A megaparsec (Mpc) is equal to a million (10^6) parsecs. An object is at a distance of 1 parsec from us if its parallax is 1 second of arc.

Periapsis - The point in the orbit closest to the planet.

Periastron - The point of closest approach of two stars, as in a binary star orbit.

Perigee- The point in the orbit closest to the Earth.

Space Science Glossary

Perihelion - The point in its orbit where a planet is closest to the Sun. when referring to objects orbiting the Earth the term perigee is used; the term periapsis is used for orbits around other bodies. (opposite of aphelion)

Planet- A large round object in space, such as Earth, that travels around the sun or another star.

Planetary nebula - A shell of gas ejected from, and expanding about, a certain kind of extremely hot star.

Plasma - A low-density gas in which the individual atoms are ionized (and therefore charged), even though the total number of positive and negative charges is equal, maintaining an overall electrical neutrality.

Polarization - A special property of light; light has three properties, brightness, color and polarization. Polarization is a condition in which the planes of vibration of the various rays in a light beam are at least partially aligned.

Pole Star - The name of the star that lies almost directly above the North Pole, which is the most northern place on planet Earth.

Positron- The antiparticle to the electron. The positron has most of the same characteristics as an electron except it is positively charged.

Prominence- a huge tongue or loop of gas that rises high above the sun's surface.

Proton - A particle commonly found in the nucleus of atoms with a positive charge.

Protostar- Very dense regions (or cores) of molecular clouds where stars are in the process of forming.

Pulsar- A rotating neutron star which generates regular pulses of radiation. Pulsars were discovered by observations at radio wavelengths but have since been observed at optical, X-ray, and gamma-ray energies.

Q

Quasar - A specific type of quasi-stellar source.

R

Radial velocity- The speed at which an object is moving away or toward an observer. By observing spectral lines, astronomers can determine how fast objects are moving away from or toward us; however, these spectral lines cannot be used to measure how fast the objects are moving across the sky.

Radian; rad - The supplementary SI unit of angular measure, defined as the central angle of a circle whose subtended arc is equal to the radius of the circle.

Radiation- Energy radiated in the form of waves or particles; photons.

Radiation belt- Regions of charged particles in a magnetosphere.

Radio Electromagnetic Radiation- Has the lowest frequency, the longest wavelength, and is produced by charged particles moving back and forth; the atmosphere of the Earth is transparent to radio waves with wavelengths from a few millimeters to about twenty meters.

Redshift- An apparent shift toward longer wavelengths of spectral lines in the radiation emitted by an object caused by the emitting object moving away from the observer. See also Doppler effect.

Reflection law- For a wavefront intersecting a reflecting surface, the angle of incidence is equal to the angle of reflection, in the same plane defined by the ray of incidence and the normal.

Relativity principle- The principle, employed by Einstein's relativity theories, that the laws of physics are the same, at least locally, in all coordinate frames. This principle, along with the principle of the constancy of the speed of light, constitutes the founding principles of special relativity.

Resolution (spatial)- In astronomy, the ability of a telescope to differentiate between two objects in the sky which are separated by a small angular distance. The closer two objects can be while still allowing the telescope to see them as two distinct objects, the higher the resolution of the telescope.

Resolution (spectral or frequency)- Similar to spatial resolution except that it applies to frequency, spectral resolution is the ability of the telescope to differentiate two light signals which differ in frequency by a small amount. The closer the two signals are in frequency while still allowing the telescope to separate them as two distinct components, the higher the spectral resolution of the telescope.

Resonance- A relationship in which the orbital period of one body is related to that of another by a simple integer fraction, such as 1/2, 2/3, 3/5.

Retrograde- The rotation or orbital motion of an object in a clockwise direction when viewed from the north pole of the ecliptic; moving in the opposite sense from the great majority of solar system bodies.

S

Satellite - A body that revolves around a larger body.

Sensitivity- A measure of how bright objects need to be in order for that telescope to detect these objects. A highly sensitive telescope can detect dim objects, while a telescope with low sensitivity can detect only bright ones.

Seyfert galaxy- A spiral galaxy whose nucleus shows bright emission lines; one of a class of galaxies first described by C. Seyfert

Solar flares- Violent eruptions of gas on the Sun's surface.

Solar mass- A unit of mass equivalent to the mass of the Sun. 1 solar mass = 1 Msun = 2×10^{33} grams.

Spectrometer- The instrument connected to a telescope that separates the light signals into different frequencies, producing a spectrum. A Diversive Spectrometer is like a prism. It scatters the X-rays of different energies to different places. We measure the energy by noting where the X-rays go. A Non-Dispersive Spectrometer measures the energy directly.

Spectroscopy - The study of spectral lines from different atoms and molecules. Spectroscopy is an important part of studying the chemistry that goes on in stars and in interstellar clouds.

Spectrum (plural: spectra) - A plot of the intensity of light at different frequencies. Or the distribution of wavelengths and frequencies.

Speed of light (in vacuo) - The speed at which electromagnetic radiation propagates in a vacuum; it is defined as 299,792,458 m/s (186,000 miles/second). Einstein's Theory of Relativity implies that nothing can go faster than the speed of light.

Star- A large ball of gas that creates and emits its own radiation.

Star cluster- A bunch of stars (ranging in number from a few to hundreds of thousands) which are bound to each other by their mutual gravitational attraction.

Sunspots - Cooler (and thus darker) regions on the sun where the magnetic field loops up out of the solar surface.

Synchrotron radiation- Electromagnetic radiation given off when very high energy electrons encounter magnetic fields.

T

U

Ultraviolet Electromagnetic radiation at wavelengths shorter than the violet end of visible light; the atmosphere of the Earth effectively blocks the transmission of most ultraviolet light.

Universe Everything in space is part of the universe. Scientists think it was formed 10 to 15 billions years ago with a big explosion. When the universe cooled down, huge swirls of dust and gas clung together to form galaxies.

Universal constant of gravitation G- The constant of proportionality in Newton's law of universal gravitation and which plays an analogous role in A. Einstein's general relativity. It is equal to 6.664×10^{-11} newtons per square meter per kilogram squared (see scientific notation).

Space Science Glossary

V

The Venera satellite series - The Venera satellites were a series of probes (fly-bys and landers) sent by the Soviet Union to the planet Venus. Several Venera satellites carried high-energy astrophysics detectors.

Visible- Electromagnetic radiation at wavelengths which the human eye can see. We perceive this radiation as colors ranging from red (longer wavelengths; ~ 700 nanometers) to violet (shorter wavelengths; ~400 nanometers.)

W

Wave-particle duality- The principle of quantum mechanics which implies that light (and, indeed, all other subatomic particles) sometimes act like a wave, and sometimes act like a particle, depending on the experiment you are performing. For instance, low frequency electromagnetic radiation tends to act more like a wave than a particle; high frequency electromagnetic radiation tends to act more like a particle than a wave.

Wavelength- A property of a wave that gives the length between two peaks of the wave.

White dwarf- A star that has exhausted most or all of its nuclear fuel and has collapsed to a very small size. Typically, a white dwarf has a radius equal to about 0.01 times that of the Sun, but it has a mass roughly equal to the Sun's. This gives a white dwarf a density about 1 million times that of water!

X

X-ray Electromagnetic radiation of very short wavelength and very high-energy; X-rays have shorter wavelengths than ultraviolet light but longer wavelengths than cosmic rays.

Y

Z